REACHING FOR A NEW POTENTIAL

A Life Guide for Adult ADD

From a Fellow Traveler

REACHING FOR A NEW POTENTIAL

A Life Guide for Adult ADD

From a Fellow Traveler

By Oren Mason M.D.

Physician, Father, Patient

Edited by David Baron, MSEd, D.O.

First Edition, 2009

Copyright 2009 by Oren Mason, M.D.

All rights reserved. No part of this publication may be reproduced or transmitted in any form, or by any means electronic or mechanical, including photocopy, recording, scanned image file, or any other information retrieval and storage system, without the written permission of the publisher.

ISBN 13: 978-0-557-12353-7

For Ben, who started me on this journey.

CONTENTS

Chapter 1 TREATING ADD CHANGES LIVES

Have ADD? Ready to treat it? Then I invite you to read this book.

I have ADD too. All of us with ADD[1] share similar traits, but each story is so very different. I begin with my own story so that you will understand where this advice originates and how my life has changed.

ADD fills my life. I live with it every second and parent two sons who have it as well. Many of my friends and coworkers have it. All of my patients have it. I have read volumes to learn what can improve all of our lives. The most life-transforming things that I have learned about living with and treating ADD are presented here, with the hope that you will discover the potential you have always felt but never realized.

I knew little about ADD until adulthood.

It had never occurred to me that I had ADD, and why should it? I was a busy, 40-something family doctor working with my wife in a successful practice. We had two wonderful sons and a beautiful home on a lake in the woods. From the outside, everything appeared to be as good as life could get.

Who could have guessed that the diagnosis and treatment of my ADD would change everything? That I would soon leave my practice and begin a new one devoted to treating people with

---

[1] In this book, the term ADD is used to refer to both ADD (Attention Deficit Disorder) and ADHD (Attention Deficit/Hyperactivity Disorder). ADHD is the term used in the official diagnostic manuals, but ADD is the more common term, so I chose to stick with the more familiar usage. The book is directed to everyone with every form of ADD/ADHD.

ADD. That we would sell our home and enlarge our family to seven children. That I would feel that I was reaching my potential for the first time in my life.

I spent my early life blissfully certain that I did not have ADD.

During grade school I was not one of the naughty and disruptive kids who spent most of their early academic life shuttling between the principal's office and the "special education" room. I was a bright kid with the best grades in my class. Nobody suspected that I had a brain disorder.

During my high school and college years, my procrastination seemed excusable to me, since my grades remained good. I figured that I typically did my best work under pressure, so my procrastination could not be a "symptom", but rather a habit or a matter of personal style. The fact that I couldn't keep my thoughts on a lecture or a sermon was evidence of my energetic mental creativity. The fact that I rambled on and on in conversation was a trait I shared with my entire extended family. Every blood relative I knew did that. I was not much different than my father, his father, my uncles or anyone else at the holiday table.

There were no academic crises that concerned my instructors. I was not hyperactive, so I did not disrupt classes like the kids with Teacher Nuisance Disorder. My mother was so efficient that she kept my dad and all five of us children organized—a heroic job, since she was probably the only one of us without ADD. The inattentive form of ADD had not been described in the 1960s. No one, least of all me, had any clue that I was not living up to my potential.

Even in college and medical school, my inability to concentrate on studies never caught up with me. I was able to cram at the last minute and still get good grades. Even though I knew that there were better ways to study, learn and retain information, I didn't adopt those ways. The fact was, though, that I *couldn't* adopt them. I could only study when the heat was on, when the test was imminent.

When I tried to study ahead as we all know is best, I only daydreamed. I would try to read, but my mind was far away. I would re-read a passage and still not pay attention. Eventually, I would put the book down and do something fun, because daydreaming seemed like such a waste of time. And because I knew that the ability to cram in the information would appear a couple of hours before the test started.

People with ADD are often unaware of its impact.

It's a common myth that people with ADD never do well in school. Years ago, a psychiatrist told me that it would not be possible for someone with ADD to complete medical school. Yet I graduated with honors from medical school despite the fact that I had undiagnosed ADD. It took something much more challenging than medical school to expose my weaknesses. It took marriage and parenting.

Real life is mentally challenging at best. Taking care of my own needs as a bachelor was something that I could manage, even if not neatly or punctually. Including my wife, Chris, in my muddled life was a whole new challenge. Marriage has exciting times and routine, but I was not able to execute the routine, because it was mundane. Remembering to show up regularly and on time for work was already challenging for me. Just remembering the simple errand I had promised to do for Chris on the way home from work proved too difficult. Simple errands, for example, like picking up our child from daycare… I've routinely forgotten far more than just a few groceries.

Chris and I spent hundreds of hours and thousands of dollars in therapy trying to improve a marriage that seemed full of promise, but fell short in the day-to-day. Eventually, we arrived at a sort of peace. The marriage was good, and we made it workable, but the burden of planning, scheduling and coordinating always fell on Chris. She and I had busy medical practices, and our two active sons were welcome but demanding additions to our increasingly hectic schedules.

As the family coordination demands increased, Chris worked valiantly to keep on top of everything and everybody, but my ability to help her had maxed out long ago in much simpler days. She felt like she had three children to manage, not two, and I couldn't argue that she didn't.

The stress of parenting exposed how poorly I maintained important routines.

It became apparent that those active boys were a little more than just active. Our oldest, Ben, was diagnosed with ADD in second grade. Little brother, Paul, was diagnosed two years later. Both were challenging to parent just as surely as they were delightful. Now, treating their ADD demanded some specific parenting; it called for a consistent routine, but I could not routinely do it.

After Ben was first diagnosed, Chris and I read many books to bring ourselves up to speed on ADD. Like most physicians we knew surprisingly little about the disorder. (In truth, I had graduated from four years of medical school and four years of residency without actually diagnosing a single patient with ADD.) We both knew so little about ADD that we went scurrying to the library to learn how to be better parents to these boys.

Through our efforts to read and learn, my wife recognized similar traits in our son and me.

First, she noticed that I did not study well. We would be reading and discussing some aspect of childhood ADD, and in the midst of it she might look over and find me reading the latest *Time* magazine. Something important needed attention, but I could not attend to it.

Second, she noticed that I was terrible at follow-through. This was not a ground-shaking surprise for her. We were married for 18 years then, and I had been bad at it for all 18. She had reached some peace with her disappointment that I could not follow through out of love for her. But it truly surprised her that I also couldn't step up to the plate when our son's well-being was on the line.

Finally, she noticed that everything that she read about ADD applied not only to Ben, but to me as well. This fact had not escaped my notice either and probably explained my need to put down books about ADD and pick up almost anything else within reach—*Runner's World* magazine, overdue bills, kitchenware catalogues, anything. One night Chris asked me to read a certain chapter in one of our books so that we could discuss it. I put it down half-way through and picked up something else.

"Why won't you read that?" she asked.

"It hits too close to home," I said with rare honesty.

"Then when are you going to get diagnosed?" she countered.

The following month I submitted to four frankly boring hours of testing after which my doctor concluded that I have inattentive ADD—the kind that is often not diagnosed in childhood, because it lacks the frenetic action level that makes hyperactive ADD easy enough for the school's janitor to spot.

The doctor takes the cure.

The following day I began treatment for ADD including, at that time, a time-release stimulant. I was stunned by the effect of treatment, as were my wife and office staff. My office efficiency improved. I got home earlier with less work left undone on my desk. My driving became less hurried and less aggressive. I arrived home after work in the evening with much more "mental energy". "You're more present with the family," Chris would say.

Happily for our marriage, things that had never improved with years of effort were suddenly better. I became a better listener. "You're *listening* to me. I can't believe you haven't changed the subject yet," she said one evening. I no longer impulsively changed the subject when I became uncomfortable. I became much more a part of the running of our household and Chris' stress level dropped. Up until that point, it always seemed we couldn't get our heads above water. Suddenly, we were caught up and had extra capacity. Projects that had languished for months were suddenly done. Things that hadn't even made it to the to-do list were done too. We had time and energy to spare.

We responded the way that anyone else would. We adopted five foster children into our family.

About the same time that I was diagnosed with ADD the "Lost Boys of Sudan" began coming to the United States. The 3700 Lost Boys and 100 Lost Girls had been orphaned or separated from family at a young age and raised in refugee camps. In December of 2000, the first of these emigrating youth began to arrive. Those under 18 were placed with foster families. We became aware of a family of five teenagers—two brothers, a sister, and two first cousins—that had been initially placed in difficult circumstances and now needed a good home.

After much prayer and reflection, we took them into our family. We moved from our idyllic but small lake home into a bigger, in-town house that snugly fit all nine people. Chris "retired" from her family practice position to ensure that everyone would receive the help needed to negotiate American life, and to prevent Ben and Paul from getting lost in the process. The story of these young peoples' journeys from the tradition-based, rural culture of the fertile upper Nile wetlands to the fast-paced, data-centric American lifestyle is far more compelling than my story here, but I will leave it to them to tell someday. My part of the story was that Chris and I were raising two boys and not coping well before my ADD was diagnosed. The result of treatment was that we were able to enlarge our hearts and our home in a profound way. We became better at raising seven children—all with very unique needs—than we were before my treatment, when we had two.

Understanding ADD impacted my medical practice.

My early concept of ADD was formed more by urban myths than by all the research into ADD. I was not aware of the recent research into its genetic basis—I thought that it was a diagnosis invented to "medicalize" kids behaving badly. I believed—as many still do—that ADD is over-diagnosed, when it is actually *under-diagnosed.*[2] I was wary of Ritalin and Adderall, aware of their potential problems, but unaware of their remarkable effectiveness and safety in expert hands. My medical practice up until this point had been more informed by myths than by real knowledge of ADD and its best treatment.

When I started practice, I knew more about all kinds of rare medical conditions than this quite common one. When physicians know very little about something, we can miss it. The knowledge that treating ADD can be exciting and life-changing changed my perspective and the way I practiced.

Every family doctor takes care of many people with ADD—probably 100 or more on average, with most of them not yet diagnosed. I began seeing the ADD right in front of me, and there was a lot of it. As many of my patients started doing better, I received more and more referrals. Very soon, ADD became the favorite and largest part of my practice.

Seeing people's lives change with ADD treatment inspired me to change my career.

That is saying quite a bit, because I've always loved being a family doctor. I've delivered over a thousand babies, and have taken care of three and four generations of some wonderful families. I've watched the children grow and given them their first school physicals. I've seen people recover from miserable illnesses and nail-biting surgeries. I've been with people as they died and as their families said good-bye. Altogether, it is a privilege to be there for people in need. So it took quite a strong pull for me to leave my patients and partners in order to concentrate solely on the care of people with ADD.

But the prospect of helping other people with ADD like me was too exciting to miss. I turned my patients over to my very capable partners and began a new practice devoted exclusively

[2] The number of diagnosed cases of ADD is rising due to an increase in awareness and the expansion of the criteria for diagnosis to include the non-hyperactive form. There is no evidence that the actual rate of ADD is rising.

to patients with ADD and similar disorders. My new office is small and homey, not big and bustling like the last. But it has been a wonderful decision; I am happier than I have ever been professionally. And I am exceedingly grateful to the patients who have shared their journeys of hopes renewed and potential rediscovered.

Part 1

First Steps

Chapter 2 THE MOST EFFECTIVE ADD TREATMENT

After diagnosis, the very first thing I encourage is a trial of medications.

It's what I did for myself, and I believe this is where you should start.

I have yet to find a book on ADD that begins with this advice. If most of your knowledge about ADD has come from popular media, you might conclude that medications are the *last* treatment you should try for ADD, and only if all else fails. Many books tout drug-free treatments for ADD. And if you've spent any time on the Internet researching ADD, you have no doubt found many people who strongly oppose medications. Many of the experts on TV talk shows tell stories of individuals who improved using their drug-free methods. You might notice that many of them have a "doctor-tested" product to sell you such as vitamins, organizers or diet plans. Whom do you believe? How is a person to choose?

No single step we can take will make as many positive changes in our lives as medication.

A large body of scientific evidence has repeatedly demonstrated this fact. Medications do amazing things for the vast majority of people with ADD. Time and time again, people who took medications had more organization, more focus and less restlessness than people who took sugar pills. It happened whether drug companies, the government or private organizations financed the studies. It happened when they tested the medications at Harvard, Duke, USC and the University of Illinois. It happened when they tested the medications in the US, Brazil, Kenya and Japan. It happened when people knew what they were taking and when they didn't. It happened for children and adults with ADD. It happened for people of every sex, race, color, ethnicity and religious preference. It happened for both NASCAR fans and NPR contributors. It happened for tens of thousands of people enrolled in all these trials.

The National Library of Medicine lists over 900 clinical studies in humans with ADD which show the improvements that medicines make. There are hundreds more studies in the laboratory that help us understand how these medicines work. ADD medications dramatically and consistently help people with ADD. Evidence is pretty good that those medications will help you too.

Furthermore, medications generally improve all ADD symptoms, whereas other treatments help in more limited ways. Everyone properly diagnosed with ADD has at least six symptoms of the disorder; some of us have all 18[3]. And those are just the official symptoms. There are dozens more symptoms that we ascribe to ADD that are not part of the official diagnosis. Medications address *all* the symptoms, not just one or two or a few. Non-medical therapies usually have a

[3] In adults, the inattentive symptoms include 1) trouble maintaining attention to detail 2) making careless mistakes 3) trouble organizing complex tasks 4) avoiding mentally challenging (boring) tasks 5) leaving projects unfinished 6) difficulty remembering appointments and obligations 7) trouble listening in conversation 8) misplacing or losing important items 9) easily distracted by surroundings or even by our own thoughts.

The hyperactive/impulsive symptoms are 1)fidgeting or squirming 2) difficulty staying seated for long periods 3) overly active or compelled to be in motion 4) feeling restless or fidgety 5) difficulty relaxing 6) overly talkative 7) blurting out private thoughts or finishing other's sentences 8) trouble waiting turns 9) interrupting others.

Again, these are considered core symptoms for diagnosis, but there are many other symptoms of ADD such as chronic lateness, poor time management, tendency to hyperfocus and emotional impulsivity. Since this book is written for people who are already diagnosed with ADD, I didn't want to spend a lot of time discussing the diagnostic criteria.

narrower effect. A smart phone scheduler with alarms can help with a habit of chronic lateness, but it won't help fidgetiness, daydreaming or impulsively bad decision-making.

ADD medications lead to impressive, life-changing improvements.

ADD is not just about misplacing car keys and being chronically late for meetings. While these are common problems for most people with ADD, they are some of the most trivial. Most of the adults I have known with ADD will express sadness or frustration over a life-long pattern of failing to live up to their potential. Our self-esteem suffers, and we are more likely than not to struggle with depression or anxiety. Our relationships suffer and careers sputter or fail. We are more likely to be divorced, to have been fired, or to have financial or legal troubles. We are more likely to use and depend on alcohol and illegal drugs. We are more likely to have said or done truly regrettable things, and we live with those regrets daily. And medications start to change these patterns.

When I was first diagnosed with ADD and began treatment, I expected that I would have less trouble organizing work, staying on time and getting paperwork done. And that happened just as expected. However, I did not expect that my relationships would improve—that I would suddenly listen better to my wife or remember the small details of my kids' needs.

I had no clue that my capacity for family and love would grow, that I would be able to start a new practice, or that I would *ever* be able to sit down and write a book. These are the real miracles of ADD treatment for me. What might they be for you? The things that get better with treatment not only make our own lives better, but can dramatically improve things for the people around us as well.

It makes sense that medications would help ADD, because it is a brain-based disability.

ADD is inherited; it's in our genes. We are born with it, and we will die with it. That means that ADD is *not* something that happens for lack of discipline (neither ours nor our parents), nor is it caused by laziness, low IQ, not trying or not caring. It is *not* caused by too much TV and video games or the fast pace of modern life. It is not caused by eating too much sugar or not enough vegetables. It is not even caused by any illegal substances we may have used in our youth.

ADD is well known to be caused by diminished activity of the neurotransmitters dopamine and norepinephrine in the dorsal-lateral, prefrontal cortex and the dorsal aspect of the anterior cingulate gyrus as well as decreased dopaminergic activity in the frontal-striatal pathways including the limbic system structures and even as far as the cerebellum.

Oops, sorry. Here I am trying to make sense of ADD and I let that gobbledy-gook out. Truth be told, I just wanted to show you that I've been to medical school and can talk like other medical experts. Let's try that again…

ADD is caused when certain things that happen well in most people's brains happen badly in ours.

For example, most people can block interference well. That means that when they concentrate on something, they are able to screen out life's distractions and focus on the task at hand. The brain is wired to do this, and it uses the places I just mentioned to do that task. But if your frontal-striatal pathways were born on the weak side, you can't do that as well. Just like a schoolmate who was born with weak muscles in one leg could never climb stairs as fast as you. He can do what you do, but not as well. And he gets tired faster.

But come time for a math test, you are the one mentally limping. He is ignoring everything happening in the universe so that he can work out those complex calculations. Meanwhile you can't ignore the individual dust molecules floating between you and the test paper much less major annoyances like the clock ticking and the fluorescent lights humming. You can screen out some distractions for a while when you have to, but your brain will fatigue pretty fast from doing too much of that. Planning and organization use the same mental muscles as math tests and suffer likewise when those same areas are born "weak".

When someone is born without a function that most people have, we call it a disability. Someone who can't hear has hearing impairment. She is *not* too lazy to listen. Someone born without nerves to control the legs is disabled, at least as far as the ability to ambulate is concerned. Someone born without normal function in any particular part of the brain is disabled, not lazy.

People who have normal function in the parts of the brain that ADD affects are better at important but boring tasks than we are. They can motivate themselves to do things simply because those things are important. When they tell us to "Try harder," they are telling us to use the part of the brain that does mundane things for the sake of importance, and engage "task persistence" so that we continue that effort.

They don't understand that we have weak "try harder" mechanisms and don't really have the appropriate brain machinery to do what they recommend. This is not because we lack understanding or desire. It is because we don't have the mental machinery to sustain tedious tasks when there is no emotional intensity to motivate us.

ADD medications help our brains function more like "normal" people's brains do.

Medications get into the poorly functioning areas of the brain and boost their function. They raise levels of the chemical messengers used by the brain to communicate between nerve cells so that our brains work much more like normal brains. The attention pathways become stronger. The ability to screen out distractions improves and it takes less "mental energy" to get things done. The ability to do important but uninteresting tasks increases.

With medication, we can accomplish a wider range of tasks and function, not normally, but at least closer to normal. It is almost like putting a brace on the weak leg of our friend. If it is a good one, he will walk more easily, with less limp, and he won't fatigue as fast. He'll get around a lot like us. He'll seem pretty normal.

When people with ADD finally experience the effects of medication, they are often stunned at the improvement.

It is wonderful to hear patients newly treated for ADD trying to describe their experience. Many will say something like, "It's amazing—I can tell myself to do a task, and it just happens. I don't have to wrestle with myself to make things happen—I'm just doing them. I had no idea that other people were getting things done with so little effort."

Many will try to express how much happier they are. Some will tear up as they speculate about how different their lives might have been had they been treated at a younger age. Others marvel at how many more life and career options they now have. Many will express surprise at finding a wonderful side of themselves that they didn't know existed. Virtually everyone reports that he or she is getting more done and doing it more easily. I feel so very blessed that a big part of my job is to sit there and listen to these stories day in and day out.

I disagree with physicians and therapists who are reluctant to use medication to treat ADD.

Time and again I hear someone say that medications are the last resort for treating ADD. "Try everything else first, and if all else fails, then and only then should you consider using medicine." Where does that advice come from? Can you name a single disorder in medicine where patients are advised to *avoid* the most effective treatment available? How would you like to hear a cardiologist tell you that you have a heart condition that medication can help, but he will prescribe it only after you have tried vitamins, herbs, and meditation classes, even though none of those are proven effective?

Furthermore, it is tricky defining "failure" of those treatments. How many auto accidents should you have before you say that the other treatments have failed? Is struggling with lousy college grades enough failure, or should you wait until you actually drop out of college? Does one divorce constitute a failure? Is it failure when you quit a job before they fire you, or must you actually get fired to need medication?

Those questions are almost impossible to answer. I don't know of a test that will tell those of us with ADD how far below our potential we have lived to date. I do know that medications have the potential to improve a broad range of life skills. Since there is no way to know how much they will help any individual, I believe virtually everyone with ADD should try them to know first-hand what they do.

Some of the consequences people fear from ADD medications are myths with little or no basis in fact.

David Neeleman, founder of JetBlue Airways, has ADD and was interviewed for a *60 Minutes* story on the disorder. He is a good example of someone who has compensated well for his ADD in the work environment. When he was CEO of JetBlue, he had secretaries and a whole layer of executives to get done what he didn't do well. I'll bet *his* desk wasn't messy.

In that interview, he talked about the advantages that ADD gives him—creativity, energy and "thinking outside the box"—which have been instrumental to him in his career. He declines to take medications for his ADD, because he fears losing these attributes.

However, there is no evidence that Mr. Neeleman is correct in his fears about ADD medications. Studies show that creativity and ADD are not linked. I have treated many talented, artistic, creative people who feel that medication has enhanced, not dulled their creativity. An artist with a roomful of brilliantly conceived but unfinished paintings may possess admirable creativity, but it is not effective creativity until the paintings are finished and displayed. Creativity is a great quality, but real life requires execution as well.

How would I counsel Mr. Neeleman if he were my patient? You already know what I would tell him. *Try the medicine.* What if it *doesn't* dull your creativity? What if it *helps* you remember to write down your latest brilliant idea? What if your racing mind slows down and you notice twice as many brilliant ideas? What if you find yourself able to develop even more talents? Most importantly, what if it helps you get home earlier, pay better attention to your children and love the most important people in your life better? Doesn't it seem worth just a try?

And what is the potential downside if his fears are true, and medication blunts a mental attribute he needs for his work? There is none. He can just stop the medication and his mental abilities return to their former state. In my practice, if one medication dulls any positive or sparkling part of a person's personality, I consider it an unacceptable side effect and move on to try out another medication. It is rarely an issue.

Unfortunately, his remarks were heard by millions of people. They might be influenced to believe that ADD medications do things that they don't do and have dangers that they don't have.

But aren't there positive attributes to ADD? Won't medication diminish those as well?

That is exactly what several celebrities with ADD and even some ADD experts argue. "ADD is an advantage if you can just harness the positive energy," or "if you put it in a favorable situation" are common themes. This is an approach some ADD coaches use to motivate clients and therapists use to address a patient's poor self-esteem. "ADD has its downside in certain situations, but be grateful for the creativity and out-of-the-box thinking it gives; those will be the drivers of your success," is one version of this approach. Of course, it is advanced by gifted people like Mr. Neeleman who have succeeded despite ADD, often in business, entertainment or medical fields[4]. The implication is that ADD has positive

---

[4] It is also a very common approach used in ads for unproven ADD therapies.

attributes that can a) lead to success if properly managed and b) that will disappear on medication. Both implications are incorrect.

Let's run through a list of ADD symptoms and pick out the ones that have been a big help to us so far. The list includes forgetfulness, trouble listening, procrastination, failing to finish things, fidgeting, blurting things out thoughtlessly and inability to relax. Which of these qualities has helped make anybody successful? Which of these would you prefer to see in a friend, an employee, a spouse? None, I venture[5].

I'm taking a controversial position here; many prominent leaders in the ADD community disagree with me. Research backs me up, though. No study has ever found something that people with ADD do better than those without. Dr. Russel Barkley, one of the most pre-eminent ADD researchers for over 30 years, recently compiled hundreds of studies in a comprehensive book, *ADHD in Adults: What the Science Says.* The book presents a sobering, detailed view of the negative impact ADD has in adults and those around them. It contains 465 pages of bad news for people who hope that there is any advantage to ADD.

There is nothing positive about ADD itself, but there are other wonderfully positive things about the people who have it.

Some—not all—are creative, some are highly intelligent, some have intense drive to succeed, and some possess brilliant intuition. According to my mom, I'm really smart, but my IQ is

[5] Hyperactivity symptoms may be the least troubling symptoms for adults. Hyperactivity is a form of energy that can be harnessed in some positive ways. But leading thinkers are beginning to question whether hyperactivity is the real issue. "Control of activity level" may be the real issue. People who are naturally active *and* have ADD are uncontrollably hyperactive. People who are naturally active without ADD are simply very active. This would also explain why so many adults with ADD are the opposite of hyperactive—sedentary and overweight but unable to increase their activity level. The inability to control the natural activity level—whether it is active, normal or underactive—may be the true hallmark of ADD.

not due to my ADD. Whatever intelligence we possess is a blessing. Mine has helped me succeed in the world of medicine despite my ADD, but it is not altered by medications. People who are intuitive or empathetic despite ADD should not think that their intuition or empathy is due to their ADD, nor that it will disappear when their ADD is treated.

Currently, it is the custom to refer to people with ADD as "ADD'ers". I don't follow that custom, because it feels wrong to me to refer to people who have many strengths and some weakness by this one glaring weakness. I consider myself to be an intelligent, usually kind, mildly creative husband and family man with ADD. My son, Ben, is an extremely personable, musically multi-talented, intelligent person with ADD. My youngest son, Paul is an open-hearted, non-judgmental, deeply insightful person with ADD. None of us is mostly ADD, even though every bit of who we are is affected by ADD.

Our mental gifts have cushioned the impact ADD might have had in our lives.

Our mental strengths are responsible for whatever successes we have achieved. But successful compensation for ADD in one area of our lives does not mean we will therefore find success in other areas. Getting through medical school with untreated ADD didn't make me an attentive driver, and it certainly didn't make me successful in my marriage or parenting tasks.

I believe it is wrong to dismiss the need for ADD medications simply because we have cleared a tough hurdle without using those medications on some occasions or in some areas. Yes, I managed to clear the medical school hurdle without the medications, but I would love to have had the mental acuity I now enjoy during those medical education years. They would have gone better, and I would be a better doctor as a result.

Every person with ADD has some wonderful attributes. I spend most of my life with ADD people and couldn't stay sane without the ability to find and enjoy the good in them. But you'll never catch me saying that ADD is a positive attribute or wishing that my non-ADD friends could have it too. I'm neither that unkind nor that dumb. Just ask my mom.

One more thing we know from all these studies: ADD medications are safe.

The problems that ADD medications cause are quite small compared to the problems they solve. It is exceedingly rare to find a case of someone who has been permanently harmed by an ADD drug (taken under a doctor's direction.) Any problems they cause clear up when you stop the drug. They are relatively minor and most people who suffer side effects from these medications consider them a small price to pay compared to the benefits they receive.

Now the concept of a "safe" medication calls for an explanation. If we say a medicine is safe, it does not mean that it will never, ever cause a problem. We consider them to be safe if the chance of harm is much lower than the chance of your original condition harming you more.

If you have a cancer that is always fatal, you might consider taking a cancer drug where 90% of the people who take it are cured but the other 10% develop fatal side effects. Most people would tolerate the danger of such a drug in view of its benefits. But you wouldn't think of taking a hay fever medicine with the same risks. A hay fever medication should not ever have fatal side effects, because hay fever itself is never fatal. You need a treatment with a lower risk than the original condition. The ADD medications we will be discussing are all far safer than ADD itself with one crucial caveat…

ADD medications are only safe if carefully used under a doctor's direction.

Knowledgeable physicians can help select and fine tune the treatments on a case-by-case basis. They monitor for side effects and help us minimize or deal with them. They monitor for adverse reactions and guide us through those rare events. The entire concept of medication safety assumes proper medical guidance, just as the concept of commercial airline safety assumes skilled pilots—not random passengers—are flying the planes.

Sadly, it is true that people who abuse ADD medications have been harmed. Every year there are a small number of fatalities in individuals who take high doses of stimulants, typically in a party-type situation and usually in combination with other illegal drugs or alcohol. Permanent harm can and will come from the misuse of almost any drug, but rarely if at all from the proper use of ADD medications.

Fortunately, ADD medications meet a very high standard of safety.

Relative to the routine and daily risks that we all face, the medications for ADD are remarkably safe. They are not simply safer than untreated ADD; they are probably as safe or safer than everyday life.

Everyday life can kill you. About six children die each year from fragments of burst balloons which block their airways. Every year or two a vending machine falls over and kills someone. Bike riding is an activity that physicians and health experts often encourage for its health benefits, yet 800 cyclists in the US die each year from accidents and thousands more sustain permanent injuries. Lightning strikes people, dogs suddenly attack, toxins creep into food and water supplies, bridges collapse and planes fall from the sky. Really unfortunate stuff happens, and we live with these risks every day.

However, there is no evidence that any current ADD medication causes any fatal problem to happen more frequently than its natural occurrence. If a medicine doesn't raise the background risk of problems, it is considered very safe. As far as we know, the trip to the pharmacy—whether by car or by bike—carries more risk of death and injury than the ADD medicine. Even the over-the-counter medications that you can buy without prescription for cold symptoms and headaches have small but known risks of harm and death. Very few medications have been examined as closely as these and found to have such low risk.

ADD is dangerous. It needs effective treatments.

It is usually right to accept some risk in our treatments if they reduce a much larger risk of the underlying disease. And this is the important point to remember about ADD: it is a dangerous condition. People with ADD suffer many more problems than people without ADD.

We have more than twice as many auto accidents, and we are more than twice as likely to be seriously injured as people without ADD. We smoke, overuse alcohol and abuse other illegal drugs at 2-3 times the rate of normal folk. If we get chronic diseases like diabetes or heart disease, we are more likely to forget medicines, run out of prescriptions, skip doctors' appointments or fail to do what we need for ourselves. We contract sexually-transmitted diseases at a rate much higher than the regular population. Crisis pregnancies are 10 times as common for women with ADD.

Nobody has scientifically studied the death rates in untreated ADD, but accident statistics[6] suggest that ADD drivers cause hundreds or thousands of highway accidental deaths each year above and beyond the background rate. In the face of all the misery that ADD brings, I believe that you should worry pretty hard about *not* treating it. Even if future research shows minor risks with the ADD treatments, the substantial risks of ADD itself must still be addressed.

---

[6] This is my own statistical projection. It's an educated guess, but *not* a fact. Research (J Clin Psychiatry. 2006 Apr;67(4):567-74) demonstrated that ADD drivers had 4 times as many accidents as non-ADD drivers and that the accidents were *more severe* on average. If 5% of the population (the rate of adult ADD) had 4 times the baseline accident rate, they caused 13% of all traffic fatalities or 5,330 of the 41,000 traffic deaths in 2007. That 5% of the population would have been expected to cause 2050 deaths, so 3280 deaths were possibly attributable to ADD's effect on drivers. Likewise, the excess injury rate would be about 200,000/yr.

But aren't these ADD medications really powerful?

I have heard and read many times that the medications for ADD are "powerful" drugs. That is a strange description for a medication. It does not refer to any scientific measurement that I know. There is no comparison that pharmacologists use to measure the "power" of a drug. As an example, consider this news headline: "Legislator Questions Use of Powerful, Mind-Altering Drugs In Children." That seems to imply that it might be better if the medications we used were weak.

When scientists study a medication, the focus of the research is on its *effectiveness*. The concept of effectiveness refers to how often and how reliably a medicine does what it is supposed to do. If you are going to bother with any medications, don't use a 'weak" one—whatever that is. But *do* use an effective one. Fortunately, the medications used for ADD are both highly effective and relatively low in adverse events.

Well then, aren't these ADD medications mind-altering?

The phrase "mind-altering" is often used in the lay press to describe ADD medications as well as anti-depressants, anti-anxiety agents and other psychiatric medications. It implies that something really nasty and permanent will happen to someone who takes them. Maybe our brains will start vibrating and shake completely loose or melt and run out our ears in the middle of the night. Okay, so we know that won't happen, but what about people who took "mind-altering drugs" in the 1960's and subsequently suffered hallucinations, seizures and changes in their political party?

It is good to raise questions about long-term consequences of drugs that affect the mind, the heart or any vital part of the body. Happily, though, the long-term effects we find with ADD medications are simply not dangerous.

For example, some have worried that introducing children to stimulants would increase their use of illegal drugs when they reach adolescence. Fortunately, this doesn't happen.

Adolescents and young adults with ADD are more likely to use illegal drugs than their normal counterparts. However, if they have been treated with medications (like Ritalin and others that we will soon discuss), there is no increase in illegal drug use. A few studies suggest that they may even help *prevent* some illegal drug use. The long term effect is certainly not negative.

Similarly, people with ADD who drive when taking ADD medications perform *better* than when they are not medicated. Anyone who shares the road with ADD drivers probably prefers that kind of mind alteration.

ADD medications may even normalize the brains of people with ADD.

Dr. George Bush from Harvard University uses MRI scans to study the brains of people with ADD when he is not busy explaining that his name is for real. In 1999, he demonstrated functional differences in the brains of people with ADD and normal subjects. Dr. Nora Volkow at the National Institute of Drug Abuse followed up his findings by scanning ADD subjects before and after taking Ritalin. She found that the brains of the ADD people started functioning like the brains of normal subjects *after* they took the medication.

Clearly, we need more studies of the long-term effects of ADD treatments, especially in adults. ADD has only recently been considered a disorder that continues into adulthood, so the majority of studies are in children. While the safety and effectiveness of ADD treatments have been studied very well, we still need to know a lot more to help us devise even better long-term treatment plans.

Types of treatments for ADD

A large body of scientific evidence can be examined to teach us what is known about other ADD treatments as well. In general, non-drug therapies for ADD tend to have narrower effects and are backed by less research. I think of therapies as falling into one of three categories.

In the first category are therapies that correct the abnormalities or normalize the differences in the ADD brain. Medications are the only treatment in this category.

In the second category are those therapies that enhance the function of the brain as a whole. They improve both the impaired and non-impaired areas of the brain. It's similar to someone with painful spinal arthritis getting a back massage. It doesn't directly affect the root of the problem, but it helps you feel better and stand straighter. It eases symptoms and improves quality of life, but does nothing to affect the progression of the underlying arthritis.

Exercise is a great example of this kind of treatment in folks with ADD. It improves not only attention and executive function but decreases anxiety and depression, deepens sleep and improves overall health. Its broad effects on the brain really help people with ADD and those without. Exercise doesn't cure ADD, but it does great things for brains that have it, so it lessens the impact of ADD.

The third category of ADD treatments involves "work-arounds", also known as compensations, accommodations or coping skills. They don't treat the impaired area of the brain, but rather recruit other competent brain systems or else external systems (technology, other people) to do the work that would normally be performed by the impaired areas. Compensations use less efficient pathways, but are often very workable. Most accommodations address a single difficulty.

Let's take a look at alternate ADD treatments in light of these distinctions. Since medications are the only example of the first category—treatments that directly address the abnormality—we will go on to the second category: treatments that are good for the whole brain of people with ADD.

Type 2—Diet and nutritional therapies

Diet is widely considered to be an effective treatment for ADD. However, the National Library of Medicine lists only six studies that have been done with various diets on ADD patients. Some diets decrease hyperactivity in a small fraction of children, but they don't improve attention or decrease impulsivity. On the whole, there is little evidence that diet helps children and no evidence at all that it helps adults with ADD. As a physician, I would not recommend diet therapy, because I have no confidence based on scientific evidence that it would help anyone's ADD. Keep in mind that I am *not* saying that diet never helps ADD. The evidence doesn't tell us that either.

Nutritional supplements have been disappointing, too. They are very popular with people who sell natural cures—as if crushing plants-you-never-heard-of into pill form and taking large amounts of them were "natural". At the moment, no nutritional supplement is known to treat ADD. A few that contain caffeine have been shown to improve alertness and test-taking. Taking supplements of omega-3 free fatty acids is a current trend. A couple studies show some mild improvements, but several others found no benefits[7]. Only a few small studies are available[8]. Every year a few "fad treatments" are popular in help forums. But one by one they fall by the wayside when subjected to quality research.

Finally, most nutritional supplements have little or no safety testing behind them. The FDA requires extensive safety testing for medications, but none for nutritional supplements or non-medication therapies. The ratio of effect-to-potential-danger may be much worse for "natural" treatments than prescription drugs, but you and I can't even evaluate the risk, because it is usually not known.

---

[7] Omega-3 FFA effects on schizophrenia and mood disorder are a different story. There is a consistent trend toward mild effectiveness in the studies that have been done so far in depressed people.

[8] I would like to see 8-10 studies involving several hundred subjects before recommending any therapy. There are hundreds of studies backing the safety and efficacy of the medications discussed later in this book. So if someone says that something or other helps ADD, there should be a substantial number of studies performed by different researchers that give consistent results.

It is reasonable to try "natural" supplements, but I would try proven medications first and do supplement trials later. Maybe omega-3 will be the next natural miracle, but I'm not buying fish oil futures for my retirement fund[9].

Type 2—Neuro-feedback

Neuro-feedback involves connecting electrodes to the scalp so that brain wave activity can be monitored, then conducting training sessions that reward the subject for maintaining a state of relaxed concentration. It has helped some people substantially, some people a little and many not at all. It can be costly and time-consuming, and it is not known how long it lasts or how often repeat treatments should be performed.

There is one published clinical trial that shows positive effects for neuro-feedback in childhood ADD but no good studies have been done in adults. Most studies of neuro-feedback are not well done; they are not randomized, and there is no control group. I have numerous patients who have tried it. Most found no lasting benefit, although some described mild or non-specific benefits[10]. Several felt less anxious, and that is truly a good thing, but anxiety is not an ADD symptom. It also provides benefits to many people who don't have ADD. There are many cases of people with depression, anxiety and memory impairments who show improvements. It is used by athletes and "normal" people to improve focus and awareness.

The effects of neuro-feedback so far seem to be broad, non-specific and unreliable. Maybe that will change as more is learned and techniques are improved. I suspect that neuro-feedback is generally good for many brains, but not a specific treatment for adult ADD. I don't tell people not to try it, but

---

[9] Actually, I'm not buying pharmaceutical futures, either. For ethical reasons, I don't invest in any companies that supply ADD treatments other than my own practice.

[10] To be fair, if people have tried neurofeedback and their ADD was cured, they won't be patients in my practice. The experience of my patients is not a substitute for careful research. Lacking such research, we have to make do with these observations which is why I offer them.

neither is there enough evidence to recommend it, especially in light of the substantial time commitment and costs. There is certainly no evidence that it can be used in place of medications.

Type 2—Working memory training

Working memory training is easier to understand than neuro-feedback. Basically, you and a computer do memory exercises. Happily, it works. Exercising your working memory improves your working memory. Unfortunately, it doesn't cure ADD. I think it's a good idea for any brain with or without ADD.

You can pay a lot of money for a skilled therapist to administer computer-based training sessions or you can buy Left Brain-Right Brain (about $20) for Nintendo DS (about $130 if your kids don't already have one) and administer your own sessions. No careful study has shown that the expensive sessions help more than doing math problems or concentration-memory games. You actually improve your working memory by working crossword puzzles, Sudoku and the like. You can even improve working memory by reading books on subject matter that is new to you. You're probably improving your working memory right now!

Other Type 2 brain-improving treatments

We've already mentioned that exercise is good for a brain, and therefore can lessen the severity of ADD. Meditation and yoga are probably similar. Both have been found to decrease some ADD symptoms, but the effect is temporary. Even if its effects on ADD are partial, who could argue with better flexibility, strength, balance and inner tranquility?

Abstinence from things that worsen brain function also falls into this category. Moderation or elimination of alcohol is good for brain function and frees many brain cells to function at a higher level than when they are burdened by alcohol's effects.

Type 3—Accommodations—finding tools to cope with ADD

There is very little research regarding lifestyle accommodations. There are literally thousands of them, and they can vary from quite simple (having your spouse write a shopping list on your forearm where you can't misplace it) to the most complex (finding someone who loves you, wants to marry you and is willing to make lists on your arm.) The hallmark of an accommodation is that it represents a way to get a task done that people without ADD accomplish using their executive centers but that we need outside help to do. Most people with ADD are adept at many accommodations and have used them for years to deal with the forgetfulness and lack of focus. Accommodations are somewhat personal, though; one that works well for me might fail in your hands or vice versa.

We can buy list systems, day planners, multi-alarm watches and color-coded home office filing systems all to help us stay organized and on time. Some of us find great help in using these. Most of us have dozens of these in piles somewhere in our home offices.

I asked one patient how often she misplaced her car keys. She said she never did. I asked how she kept track of them. It turned out that she made about 10 sets of car keys so that she could always locate at least one. Many people use this same system for reading glasses. But let's say we've been misplacing our cell phone or laptop computer. It's going to take a different and better system than this.

Technology can provide some helpful accommodations. You can buy a transmitter that wirelessly pages several different color-coded smart tags[11]. You attach them to car keys, cell phones, wallets, toddlers, etc. When you misplace any of these tagged items, press the corresponding button on the transmitter unit and the smart tag starts beeping so you can locate it. Just

---

[11] They have names like Easy 2 Find and Loc8tor. I haven't use or tested them, but some patients have found them helpful.

don't lose that transmitter. Or be like the key lady and buy 8 extras.

A smartphone scheduler can store our schedule where it is convenient, and alarms can help us arrive on time. Folks with distractibility can improve concentration by wearing headphones that play "white noise" or by working in a quiet location. Those who struggle to remember to pay bills on time can set up automatic payment systems so that bills are never paid late. Those that lose track of checking account balances can set up lines of credit to back up the checking account.

Partnering with the right people can provide excellent accommodation, too. ADD adults that struggle to maintain long-term projects can hire coaches to help keep things on track and on schedule. My wife, Chris, is very organized and has kept my life from imploding a thousand times. Secretaries and assistants can turn chaos into efficiency in many cases.

We will talk about non-drug therapies for much of this book. There are many to consider, and their collective impact is huge. But let's be honest—those of us with ADD have been developing coping mechanisms ever since day one. How many different methods have we started over the years to become more organized and efficient? How many worked for a little while? And how many are still working?

Accommodations work better with the help of medication.

The problem is not that the methods themselves don't work but that we are unable to sustain the effort they require. Many times our spouses have made lists for us that we either lost or forgot to look at. Even the organizational crutches that we need require a baseline of organization that we cannot maintain. A day planner ought to help, but keeping it up-to-date so that it can remind us of an appointment takes more organization than many of us have. It is much easier to use one reliably *after* we are on medications.

In sum, there are numerous treatments available, but the one that starts to fundamentally change life for the vast majority of people with ADD is medication. Other treatments have smaller, temporary effects or don't address the actual brain differences.

Let's consider a story of a different disability.

Marlon Shirley is a remarkable person with a disability and a great website. (Check out www.marlonshirley.com) He lost the bottom part of his left leg in a lawn mower accident as a young boy at the orphanage where he was raised. He is now the fastest amputee sprinter in the world, and the only amputee to run 100 meters in under 11 seconds. He wears an odd-looking athletic prosthesis, but there is nothing odd about how fast he sprints. It's beautiful. Marlon Shirley is one of the most amazing sprinters you could ever hope to see.

But without that artificial leg, he cannot sprint. He can hop, and if you give him crutches, he can walk. That is his disability; there are just a few things he cannot do. Just so you know, there are *only* a few things that he cannot do. He interrupted his studies in aeronautical engineering to devote himself to his racing career. He has become intrigued by the high-tech design of the prosthesis he runs with, and he may eventually enter the field of prosthetic engineering. He is a motivational speaker as well. But without a prosthesis, he cannot run.

Now, a prosthesis all by itself is not much—just a lonely lump of carbon spring steel. And if you just hand a fellow like Marlon a prosthetic leg, he will not suddenly become a sprinter. That takes work, time, effort and heart. The prosthesis is not the end of his treatment, it is the beginning. Without Marlon's upper body work, endurance runs, quad presses, healthy diet, focus and training plan, the whole package still doesn't turn into a world-class sprinter. Conversely, all that training has little effect until after he acquires the artificial leg.

Medications for ADD are somewhat like Marlon Shirley's artificial leg.

Without medications, those of us with ADD simply limp in our focus and attention. We cannot begin to do with our full effort what other people do with little effort. It is silly to say that we will remember to do something, when the part in our brain that reminds us to do things is so weak. We forget and then appear lazy and uncaring. But with a little medication help that goes right to the heart of the matter (or the gray matter of the brain in this case), our mental training efforts gain real traction. Our efforts to get organized actually work, and they last more than a few days.

After we talk a bit more about the medications for ADD, we will go on and discuss all the other things that can be done to improve our ADD brains. All these remaining tasks are like the training that Marlon Shirley does. He does his training *after* the leg is fitted, not before. His training actually develops the non-disabled parts of his body—his upper body, trunk and right leg—to work *with* the new prosthetic leg. If we are to reach our potential, we, too will strengthen the non-disabled parts of our brains—our intelligence, creativity, memory and emotional selves.

It is the combination of treating the disability specifically and strengthening the non-disabled parts generally that helps us succeed. Effective treatments, together with some hard work from the parts of our brains that *do* work well, may enable us to do some pretty amazing things. Some of us will go back and complete our education. Some of us will finally feel we are reaching our potential. And miracle of miracles, some of us will now remember to pick up our kids from daycare.

Chapter 3 THE MOST EFFECTIVE TREATMENT FOR *YOU*

To determine the best medication, we will likely need to try the three major ADD medications.

When I was diagnosed with ADD and decided to begin medication, the number one question on my mind was "Which one will help me the most?" I believe that is the question everyone should personally research. But, this time, the answer does not come to us from the scientific studies. In fact, that research tells us that we will have to answer that question in a different way. We need to answer the question by completing individual trials.

The medications that treat ADD are remarkably effective when you look at how they affect large groups of people in high quality studies, but no single medication works for everyone. As a rule of thumb, the main medications for ADD work well for approximately 70% of the people who try them. And when you are in the 30% not helped by the first medication, you still have a 70% probability that the next will help you. And if you are in the 9% who don't respond to the first two medications, you *still* have a 70% chance that the third will help, and so on. Eventually, virtually everyone who persists will find help.

But the research is silent on the one question we wish it would answer—which one is best for me?

There is no body of evidence which indicates that one drug always works in women with ADD or depressed people with ADD or Swedes with ADD. All the major medications are effective for both inattentive and hyperactive symptoms. And even if it is discovered that one tends to work better in a certain group of people that includes you, alternative choices might still be better in your specific case.

But wait. Shouldn't my doctor know what's best for *me*? Isn't that precisely what doctors are supposed to know? If he doesn't know what is best for me, how am *I* supposed to know?

Great question! Medication choice is difficult, because of a unique characteristic of ADD. Since it is present from birth, it represents a lifetime of living below our *potential* function. There are not many circumstances in which someone walks into a doctor's office and says, in effect, "My life is the same as it always was, and I want you to fix that."

Most diseases represent the loss or decline of a previously normal function. Patients typically want to get back to the way they used to be before the onset of the current misery. It's easy in such cases to tell when the treatment is complete, since the misery is cured. The headache is gone, the frozen shoulder can move again or the fever has broken.

But when you ask your doctor for ADD treatment, you are not really asking to return to normal. You are asking the doctor to enable you to do things you have never been able to do; you are asking to go beyond what you know to be normal. You know you have been hampered by lack of focus, but you do not know how much your focus might improve with treatment. You hope to become less impulsive or less distracted, but exactly how much less can you expect? Since you were born with ADD, you do not have your "normal baseline" firmly in mind. If you are mentally limping, should your goal be to mentally walk? or to jog? or to sprint like Marlon Shirley?

Well, Marlon Shirley did not know how fast he could run until *after* he put on his athletic prosthesis. And you will not know how mentally adept you might become until after you try medication. And neither you nor your doctor can know which medication makes you mentally sharpest until you complete the trials.

"Try all the medications" sounds pretty daunting, but there are only three "first-line" basic medications on the market right now that are approved by the FDA for use in adult ADD and that have been widely used. There are some intriguing new ones on the horizon in the next few years, but as of 2009, we are talking about trying three medications. That is not so unreasonable.

The three medications are methylphenidate, amphetamine and atomoxetine. They are more widely known by their trade names: Ritalin[12], Adderall[13] and Strattera. We will talk about them in some detail in the next two chapters.

Another reason you want to find the best medication available for your specific case is that you will probably be on it for a long time.

In all likelihood, you are going to be on medication for many years. Since these medications don't cure ADD, we need them continuously, just as a diabetic needs insulin. So a small difference between two medications may seem unimportant if you are only thinking about the next 2 or 3 weeks, but that difference might be important if you think about it long-term.

Let's say you were initially placed on one drug and had a good response to it. In fact your response was so good, that you don't even desire to do better. Why mess with success? But what if another medication works just as well but it lasts a bit longer? Or costs a little less? Or has a bit less side effect? Over the next 5 or 10 or 20 years those little bits of improvement could add up to quite a large difference.

---

[12] Ritalin is actually just one of several brands of methylphenidate. You may be familiar with Concerta, Daytrana, Focalin, Metadate, Methylin or others. They differ in how long they last and even how well they work in one person versus another, but they all use the same main ingredient.

[13] Adderall is sold as Adderall or Adderall XR, and is so closely related to Vyvanse, Dexedrine and Dexedrine Spansules that I include them all together in the same group.

And what do you have to lose for trying? If the alternatives are disappointing, you simply return to the better treatment. There is nothing to lose and perhaps something to gain.

Let's do a little myth-busting about medications in general.

I enjoy speaking to physicians and patients all over the country on the subject of ADD. Many comments and questions come up so frequently that there must be an active myth machine circulating the misinformation. I'll repeat what I have heard, and then we'll address the myths that underlie the words:

> *"I've heard that Adderall is best for severe ADD and that Concerta is for milder cases."* Or alternatively: *"Strattera is for the inattentive type."*

First, there is no scale that measures ADD severity. Doctors and therapists may speak of "severe cases", but they are usually pointing out that an individual is really struggling in his or her life at the time[14]. My sons have been known to say of an overactive playmate that he has "A-D-H-H-H-H-H-H-H-H-D", meaning that he is impressively hyperactive. But there is no sanctioned definition of severe ADD or mild ADD or anything in between. Try to keep yourself from ranking different people's ADD.

Second, there is no ADD medication that is "stronger" or "weaker" than another. Remember that we need "effective"—not "powerful"—medicine.[15] Researchers have done very few

---

[14] While there is not an ADD severity scale, it *is* proper to rank ADD as *mildly* or *severely impairing*. People with ADD who have above average intelligence and grew up in stable families, received a good education and suffer no other mental/emotional disorders may have fewer or milder impairments as an adult. Conversely, individuals who grew up in chaotic homes without consistent discipline, who received poor or incomplete educations, whose ADD is accompanied by anxiety, depression or substance abuse disorder, or whose socio-economic circumstances are hard may face far more struggles in adulthood from the combined burden of all these disadvantages.

[15] Technically, I just glossed over the notion of "potency". Some drugs are more "potent" than another, so we use them in lower doses. 72 mg of Concerta has equal potency to 30 mg of Adderall XR. We do not consider Adderall XR to be "stronger" than Concerta simply because fewer milligrams are required to have the same effect.

studies that pit the medications against each other. Most research focuses on one drug at a time. What these studies have conclusively shown is that each of the ADD medications makes a significant improvement in most ADD symptoms in most people.

> *"Concerta has really helped my friend, so that's what I want to try."*

> Or alternately,

> *"Strattera didn't work one bit for my neighbor's son, so don't put me on that."*

I hear many variations on this theme. It is tempting to think that someone else's experience or genetic similarity will help guide our own trial, but often it doesn't. I treat more than one member of many families and there are sometimes radically different responses to medications from people who are closely related. Austin[16], an eighth-grader, had an impressive response to Strattera, but we could not find a single improvement when his younger brother, Clay, tried it. Clay responded quite well to Concerta, but when Austin tried it, he had more side effects than benefits. It's no matter. They are both successfully treated now, but not because genetics guided the selection.

In the same vein, treatment failures have many possible explanations, but other people's experiences of treatment failure are not generally relevant to our own, and they are certainly not contagious. Treatment failures may be due to genetics in some cases, but we are not able to test this in real life outside of research labs. Some people fail to respond to typical dosage trials, but then do respond to unusually high or very low doses. New drugs arrive continuously, and doctors must learn more each day about how best to use each one.

> *"I want you to put me on Strattera, because it's not a stimulant."*

When clonidine is used to treat ADD, the typical dose is 0.3 to 0.8 mg daily, so we consider it a more potent compound, but not a "stronger" one.

[16] All patient names are changed to protect identities.

If you have fears of using stimulants (or any drug category) for ADD, I hope to reassure you, because the proven benefits for most clearly outweigh the rare problems. There are disadvantages to stimulants, for sure. Some are sold illegally on college campuses for use as "study drugs". Some people who are misinformed fear becoming addicted. Addiction does not occur at the dosages recommended in ADD. Many fear the stigma of being "on Ritalin". Some don't like that stimulant prescriptions cannot be refilled, so they need a new prescription every month, and that means a lot of trips to the doctor's office in a year. And physicians dislike the inconvenience of frequently rewriting stimulant prescriptions. But this is not about your or your doctor's convenience; it's about you reaching your potential. If the medications that work best for you are inconvenient, then so be it. And if they are stimulants, so be it.

Furthermore, what does the classification (stimulant vs. non-stimulant) of a drug have to do with whether it is best for you? Classifications are artificial. There are epilepsy drugs and anti-depressants that are occasionally used to help people with ADD. They were given these classifications because the first uses we found for them were in patients with epilepsy and depression. The classifications were not given so that they would never be used for anything else. The ADD medications can likewise be used outside their classification. Strattera, for example, helps some kids who don't have ADD control their bedwetting, and researchers are testing its effects in dyslexia, anxiety, dementia and cancer patients with chemotherapy side effects.

Viagra is an interesting example of what can go awry with drug classification. It was developed to lower blood pressure, but it didn't do that very well, so it is not classified as an anti-hypertensive. Its eventual classification has more to do with a very interesting side effect that the male research subjects reported. But it turns out that Viagra can also be used to treat altitude sickness—a rapid, sometimes fatal, illness in mountain climbers. If you were a mountain climber with life-threatening altitude sickness would you want to have a discussion with the medic about the classification of the drugs he was using? ("Don't use that sex drug on me. I have a *pulmonary* condition.") I'm hoping you would confine your interest to its effectiveness. In sum, don't put too much stock in a drug's classification.

> *"I tried my friend's Adderall and it really helped a lot. That's what I want you to use for me." Or the corollary, "I tried my friend's Adderall and it felt awful, so don't put me on that."*

I have two comments here. Number one, please don't do that again. Never take other people's medicine. Different people take vastly different dosages. Some respond with great sensitivity to small doses and would be overwhelmed with normal doses. Some require staggeringly high doses that would be quite toxic to most people in order to achieve the same effect. Indiscriminately taking medications meant for another person is not proof of anything except poor judgment. Secondly, the fact that one medication works for you does not mean that another won't work better. It is fine to start with something that seems to work from your unofficial trial, but you still don't know which works best for you until you have tried a variety.

> *"My doctor says that she thinks ____________ (fill in any drug you want; I've heard them all.) is the best, and she won't let me try another."*

You probably know what I'm going to say here. It is understandable that some doctors over time might develop better success with one drug than the others. However, that is not the drug's fault. Some of the ADD medications are tricky to use properly. Physicians are busy these days, and trying all the medications will take their time and effort as well as yours. But you know better than to try only that one. You don't want what worked best for your doctor's last patient. You don't want the drug she has used the most. You want the very best you can get. Keep trying. If necessary, ask her to send you to someone who will let you try the others.

"My health plan won't pay for _____________ (fill in any drug they exclude)."

This is a tough one, but don't give up too easily here. Health plans that restrict drug choices do so out of concern for cost. We tell them that costs are paramount, because we almost always choose the least expensive plan when we get a choice, so what else do you expect them to do? But health plans do not know which drug works best for you. The way they look at, in fact, goes something like this. "Pill A works 75% of the time and costs $30 per month. Pill B works 75% of the time and costs $50 per month. Pills C and D work 75% of the time and they cost $150 per month. These medications are all the same except for the price. Let's exclude C and D, so we can save our company a lot of money."

Rather than getting all worked up and saying unkind things about health plans, let me suggest something possibly more productive. Do a trial of the medications they want you to use and a trial of the ones they don't. (Often physicians can provide samples or coupons when you try a new drug.) If you do best on a medication they exclude, write the pharmacy committee of your health plan and tell them exactly why it worked better for you than their choices. Ask them to make an exception for you. If they won't cover it in the face of evidence that it is the best treatment for you, then we'll call them all of those unkind names.

But don't be shortsighted just because they are shortsighted. If health plans will not pay for our medications, we still need the medications. Maybe it's time to check out alternative plans. There are still other ways to provide ourselves with treatments when a health plan will not[17]. But let's not allow them to make the decisions for us. If we limit our choices to the list they give us—a list that had nothing to do with our individual best

---

[17] RxAssist.org is the website of an organization that helps people find prescription cost assistance programs.

interests when it was drawn up—it is only ourselves and our loved ones that lose.

What if the very first medication I tried was miraculous? Do I have to stop it to try another?

No, at least not right away. Enjoy all the benefits, and start learning more and better ways to get organized, pay attention and improve your life. When you are very familiar and accustomed to the changes that a medication brings, it is often a simple matter to try an alternative, and it can be done at any convenient future time. Next summer, next year—either is fine.

What would someone do if no medication really helps?

Despite these glowing reports of how much ADD medications help most people, occasionally a person cannot find one that fits well. Some get good results, but have side effect problems. A few have medical contraindications to many or all of the medications. A very small number simply do not have good results with any of their trials. Fortunately, we are talking about a very small minority. There are many reasons for treatment failure that can be explored.

Those of us with ADD who also have anxiety, depression, dyslexia or bipolar disorders may need simultaneous treatment of the other problem before we will see progress in our ADD. Sometimes, a referral or second opinion can help uncover the source of a treatment failure.

The fact remains that medications are not a panacea, but simply the first, most logical, and usually most effective step that we can take. Chapters 2 to 5 of this book are devoted to medications, but chapters 6 through 14 are about non-drug therapies for ADD. In other words, there is still a lifetime of other therapies for ADD beyond pills.

Finally, I would confidently take the ADD medication(s) that work(s) best for me.

The whole point of the medication trial is to reach a conclusion and make a decision. If the medicines don't work, don't take them. I say this mostly to be thorough, because medications help virtually everyone with ADD. And if they help *you,* take them reliably so they can do their best for you. That way, down the road when somebody tells you, "You won't believe how well my brother with ADD is doing since he started taking ___________," you won't have to wonder if it might be better for you. You'll already know.

Personally, I did try all the medications. In my case, they were all miraculous.

When I was first diagnosed with ADD, there were only two first-line medications—Ritalin and Adderall. I tried a long-acting version of methylphenidate called Concerta and was stunned by the changes. It was suddenly much easier to stay organized, on task and focused. I got more done in less time, thought more about other people and had more energy. It was easier to listen to people when we talked, and to remember what they said. I was able to do mundane, simple things without the mental effort it had always taken. The changes weren't necessarily dramatic, but they improved literally hundreds of little tasks and events in every day. The cumulative effect was the miracle.

I will also tell you that the experience of the medicine per se, was not unpleasant, but neither was there anything attractive about the feeling of having Concerta in my system. It was not much different than the experience of taking a hefty dose of decongestant (such as Sudafed). I could feel my pulse rise a little, and a little bit of the same "buzz" that I usually have with a large mug of coffee. These were very minor symptoms and they diminished after a couple months. The point is that there was nothing pleasurable about the experience. The only pleasure was in the *effect* of the medication—better organization and focus.

I had some positive effects from 18 mg of Concerta, but more from the 36 mg dose, and even more from the 54 mg dose. 72 mg, however, made me feel jittery, so I decreased to 54 mg every morning and stayed on that dose for several years. Concerta can last up to 12 hours, but I had closer to 8 hours of effect from it, so I also took another dose of short-acting Ritalin at 3-4 pm to help stretch the effect of the medication into the evening hours. Altogether, I got 12 good hours of effective ADD symptom relief. The first and last hours of the day, I still struggled with my ADD .

Adderall had a very similar effect for me. If Concerta hadn't been invented, I would have taken it happily, and it would have been miraculous, too. I felt a very slight sensation of jitteriness with the Adderall that I didn't have with the Concerta, so I chose Concerta. However, I could easily have put up with that rather minor side effect if Adderall was my only effective choice.

I had been on stimulants for a few years when Strattera was introduced. I also responded well to it, so I had 3 medications to choose from. My symptom reduction was approximately the same on all three. However, the experience of around-the-clock symptom relief from Strattera seemed like another miracle on top of the first one. Stimulants work quickly with clear effects 30-45 minutes after taking them. With Strattera, there wasn't even that little bit of lag time; I could wake up and "hit the ground running". The effects of the medication didn't perceptibly change through the day or into the evening. I prefer Strattera's consistency. Its effects are present all day, through the evening, at bedtime and on arising the next day—*all* the hours that ADD symptoms have impacted my life.

It is a bit dangerous telling my individual story. What worked well for me may work differently for someone else. I'm very fortunate to be able to choose between all three medications. The larger point that I hope becomes clear is that it is worth the time, patience, trial and effort to arrive at the best regimen possible. The best one for *you*, that is.

Before we go any further, I need to issue this:

ADD RED-LEVEL TERROR ALERT!

Chapter 4 is full of FACTS and DETAILS! Even worse, SO IS THE NEXT! There's a pretty good chance that you will find them—uh-oh—BORING. Prepare yourself. Most of the ADD people I know would rather face pain, hunger and intolerable physical conditions than boredom. Truth is, these chapters are in here to be complete and thorough. If you are not at a place in your ADD journey where you need to know the details of medications and the differences among them, you may want to skip ahead to Chapter 6.

Chapter 4 OVERVIEW OF ADD MEDICATIONS

As we have discussed, three medications have emerged that have excellent efficacy and safety. More are on the horizon. Some understanding of the differences among these medications will assist us as we evaluate them. It is traditional to begin with the oldest and best known medications first and end with the newer ones. I've changed things a bit in this book and listed them in the order that the FDA approved them for use in adults. It's a personal quirk; I just can't do it the same as everyone else.

Strattera has made a miraculous difference for many adults with ADD. I would hate for you to miss the chance to try it.

The generic name for Strattera is atomoxetine. It was the first drug approved by the FDA for the treatment of ADD in adults. Strattera was developed in the 1980's by the researchers at Eli Lilly who discovered fluoxetine—better known as Prozac,—but it was not tested for ADD patients until the late 1990's.

Strattera was new and different in many ways. Especially important for our purpose now was that scientists tested Strattera in adults to show that it is safe and effective for us as well as children. Before Strattera, no pharmaceutical manufacturer had presented data on adults to the FDA. Many small studies of stimulants had shown them effective in adults and so they were already in wide use by physicians for adults with ADD. But the studies that Strattera's manufacturer presented to the FDA tested many more people over a much greater time span than *all* the previously published studies on stimulants in adults combined. In December of 2002, Strattera was approved by the FDA and released to the public.

Strattera—how does it work?

Strattera is quite different from the stimulants; it is classified as a "specific norepinephrine reuptake inhibitor" (SNRI). See what happens when they ask researchers to classify things? I could have come up with something much more descriptive like "nuclear brain focuser" if they had just asked.

An SNRI works by raising levels of norepinephrine directly and dopamine indirectly in the parts of the brain that plan, organize, remember and execute details. People with ADD have lower levels of these two chemical messengers. This helps the different parts of the brain "converse" more easily with each other. The part of your brain that is planning the route home from work needs to be in contact with the part of your brain that remembers what (or whom) you promised to pick up on your way.

Strattera's effects are not immediate. Careful observations can detect the beginning of effect in less than 24 hours, but the effect is too subtle to bother about. Remember, we are seeking large effects. The effect continues to build so that is noticeable in a few days and keeps growing impressively over a few weeks for about 60-70% of adults.

Strattera has the same end results that the stimulants do. It is effective for all forms of ADD—inattentive, hyperactive and combined. No studies have been done including Strattera and a stimulant over enough time to compare them adequately. We cannot say that Strattera works better or worse than stimulants. We can only say that it works differently, and that you will have to try it yourself to know what it does for you.

When I first recommend Strattera to patients, some are dismayed by the time it takes to see the full effects of the medication and want to try something "quicker" instead. Most of us have more than one thing going badly when we begin treatment. However, most of us always have and always will have several things ongoing that ADD treatment could improve. The medication treatments should be considered in terms of what they will do for us over the years, not what they will do this week.

Strattera's unique effect—all-the-time symptom relief

We live our ADD lives 24/7. ADD affects every decision, every mental process of every minute of every day. It is not just a problem during work hours. We drive at all hours of the day and night. Sometimes we socialize late into the evening, and sometimes we face choices that can have lifelong consequences in the wee hours of the night. There has been a tendency for many people to treat their ADD as if it were a disease of school hours and study habits or of work hours. But ADD affects every minute of every day whether we are awake or asleep.

Some of the most important hours in my own life are the first and last hours of the day. On really busy days, these couple hours might comprise the sum total of my family time. They are times when I especially want to be able to listen, control my emotions, get things done, and plan my days. Doing all that well is tough for anyone, but much tougher with untreated ADD.

Strattera is an ideal medication in this regard. It does not have a limit to its hours of effectiveness as the stimulants do. When the stimulants were our only option, we accustomed ourselves to treating ADD in specific blocks of time—the school hours, the work hours, etc. But a medication that works as well at midnight as it does at noon has a clear advantage over one that has to ramp up at the beginning of the day and wean off near the end.

Strattera's steady relief of ADD symptoms makes it tough for some people to appreciate its effectiveness, especially those used to stimulants. Stimulants have a startlingly clear onset of effect each morning, and each night users return to their baseline of significant symptoms. Every day contains those reminders of a stimulant's effectiveness. With Strattera, the symptoms improve gradually and the memory of how pervasive the symptoms used to be has to extend back several months.

Human beings are notoriously good at forgetting pain, so we quickly accustom ourselves to the new level of brain function. It is quite common in my practice for a patient who has been using Strattera for a month to report that it isn't doing very much. When we repeat the original psychological assessments, though, a different picture of Strattera's effectiveness emerges. Or better yet, if we ask the spouse or significant other, we often hear of improvements that the patient wasn't even aware of.

Strattera is new on the market, and we get better each year at understanding how to use it best. I see many patients in consultation who have tried Strattera and thought it ineffective, but it seemed to work much better the second time around. Let me encourage you not to throw in the towel too soon on this one. Around-the-clock symptom relief is an amazing option to have available for treating ADD.

Strattera—how do I take it?

The typical daily dose of Strattera for adults is 80 mg. That dose will be correct for many adults. Some larger adults need larger doses and some sensitive people need less. Some people need a little time to get used to the medicine, so it is generally begun at a lower dose and gradually raised to the target dose of 80 mg over a few weeks.

There is no need to take Strattera at a specific time for it to work properly. You can take it at night if that is most convenient. Most people take it once daily, but it can be taken twice if that works better.

It can be taken with or without food. It is my own practice to have people start Strattera with their evening meal. This decreases the abdominal pain or nausea that may occur initially. Fortunately, these are "nuisance" side effects, since they do not indicate any internal damage, and they usually disappear within a few weeks. After the first month, most people change to morning dosing with breakfast.

Strattera—what are the side effects?

Overall, about 8% of Strattera users discontinue it because of uncomfortable side effects. This is in the same ballpark as the stimulant medications that we will be discussing next. Minor side effects are those that are considered small compared to the overall benefit of the drug. Even the major ones that lead to discontinuation are usually not safety concerns.

The most common side effects with Strattera are dry mouth, nausea or abdominal pain, fatigue, trouble sleeping, constipation, urinary slowing, erectile dysfunction (in men) and decreased appetite. The only one of these that occurs more than 10% of the time is dry mouth.

Most of the side effects can be addressed by one of a few strategies. The dose can be lowered, it can be split into two smaller doses and taken ½ in the morning and ½ in the evening, it can be taken with food and the timing can be changed. If you have daytime fatigue after taking Strattera in the morning, try taking it in the evening. If it decreases your appetite, and you are one of the lucky few who doesn't need or want that, try taking it at bedtime and so forth.

Most of the side effects are quite minor. They cause discomfort but not danger. And fortunately, they generally subside with continued use. Much like climbing into a hot tub, the initial discomfort disappears as our body adapts to the new condition.

Strattera—what are the risks?

In 2004, Lilly added a warning to the Strattera product information following two cases of liver imflammation that were probably caused by Strattera[18]. Two patients developed symptoms similar to a case of infectious hepatitis. Their skin

[18] Actually, only one of the patients had a case that was "probably" linked to Strattera. The remaining patients were considered "possibly" due to Strattera. It is tougher for the FDA to evaluate adverse reactions than you might think at first. Many people who are not on drugs get similar liver problems, so it is hard to tell whether Strattera was the actual cause of the problem or just an innocent bystander.

turned yellow, they were tired and they experienced upper abdominal pain on the right side. The symptoms cleared up after Strattera was stopped, and they recovered without any long-term effects.

So far, this has happened for 6 out of the more than 6 million people who have taken Strattera. It's hard to conceive what a one-in-a-million risk looks like, so let's compare to some risks we face daily. Most of us ride in cars on a daily basis. Our lifetime chance of dying in an auto accident is about 1 in 88—1 in a million every two days. Many of us also use bathtubs regularly. Our chance of dying from injuring ourselves in bathtubs is about 1 in 11,000 mostly due to falls that cause head injuries. In the United States, our lifetime chance of being eaten by a grizzly bear is 1 in 1.2 million. That's about our chance of having a liver problem caused by Strattera—and a non-fatal liver problem at that. The greater danger of Strattera appears to be the drive to the pharmacy to pick it up. If you are really a safety conscious person, you might want to walk to the pharmacy.

Strattera's warning about suicidal thoughts.

There is no evidence that Strattera causes suicide. And there is no evidence that it causes suicidal thoughts in adults. But there *is* weak evidence that it *may* cause suicidal thoughts in children or adolescents. This is a complicated issue that is still argued among medical experts, so I've given it more prominence and detail in this book than it probably deserves.

The FDA added another warning to Strattera in 2006. This was a "black box" warning, so-called because the FDA literally requires the manufacturer to draw a black outline around this detail of the product information. The FDA determined that Strattera was associated with a 0.4% rate of suicidal thoughts in children when they reviewed the original studies done on the drug. The children who were taking placebo pills in these studies did not have any recorded suicidal thoughts[19]. Unfortunately, these were not studies of suicidal thoughts, and they don't have the scientific strength to lead to conclusions or guide our practice. The FDA took a conservative approach to this information by adding such a strong warning.

I don't disagree with the FDA and strongly value their mandate to protect us, but do want you to know how tricky it is to interpret and respond to this warning.[20] In my opinion, we would be better served if the government would support research to investigate why people with untreated ADD commit suicide three times more often than those without it, and how well treatment improves that. In that study, we could look very closely at the medications to see if any or all of them affect suicidal thoughts and behaviors in positive, negative or mixed ways.

Patients in my practice have experienced suicidal thoughts after taking all the ADD medications, but this has been exceedingly rare and has usually indicated unresolved depression. The take

---

[19] Several experts have noted that the rate of suicidal thinking in children with ADD not on medication is 10% or more. That raises the question of why such low rates were found in this study. If the study didn't detect a normal incidence of suicidal thought in the untreated group, how do we trust the measurement in the Strattera group?

[20] The information on suicidal thoughts was not the purpose of the studies, and the studies weren't really powerful enough to show what they appear to show. Researchers were not specifically looking for and carefully evaluating the pattern of suicidal thoughts in these studies. I know that sounds weird. Scientific studies are each designed to answer specific questions. If researchers notice something apart from the original intent of the study, the next step from a proper scientific standpoint is to design a new study that investigates the question raised in the original study. Pretend that you were browsing your own city on Google Earth and the satellite photo (probably a couple years old) showed a traffic jam on the route that you normally drive home. Now you are certain that your route is *sometimes* congested, but you don't how often it is, or—more specifically—if it is congested *now*. You should follow up with the right study that answers your specific need—for example, a live report from a traffic copter—before you decide whether to avoid that road tonight. The FDA could have ordered the manufacturer to do a careful study that focused precisely on the question of suicidal thoughts, but instead chose to broadcast the question raised by the first study as a strong warning. In their judgment, it was the safest thing to do.

home message is this: if you feel significantly worse and are taking Strattera or *any medication,* you need to see your doctor and get to the bottom of it.

There are some cautions that must be exercised when you take Strattera. It raises blood pressure and pulse counts by small margins in some people, and in rare cases by 10-20 points or more. Make sure you have these vital signs evaluated regularly on any medication. Stimulants can have the same effect, too. Strattera can cause palpitations (irregular heart beats) that should be investigated if they occur. No laboratory monitoring is necessary with Strattera. People with narrow angle glaucoma should not take Strattera. It is possible to develop an allergic rash or hives from Strattera, as with any drug or food. This is rare, but contact your doctor quickly if this occurs.

Adderall XR, Vyvanse, Dexedrine—Adderall has made a miraculous difference for many adults with ADD. I would hate for you to miss the chance to try it.

Adderall, Adderall XR and Vyvanse are the most commonly used versions of amphetamine (AMP). Dexedrine, in pills and slow-release capsules, was used commonly in past years, but has fallen out of favor recently. For simplicity, I'll refer to all these compounds as 'Adderall' rather than the more cumbersome generic class name, amphetamines.

Adderall has been available for more than seventy years, so we have much experience with their long-term use and safety. Adderall XR was the second drug approved by the FDA for use in adults with ADD. It is manufactured by Shire Pharmaceuticals. Shire did extensive testing with hundreds of adults prior to receiving FDA approval, so they added significantly to our evidence database for ADD medications. Their studies confirmed what other studies over many years have shown—Adderall is a relatively safe and effective treatment for both adults and children with ADD. Vyvanse was released by Shire with the same indications in 2007.

Adderall belongs to a larger class of medications called stimulants. Methylphenidate (the active ingredient in Ritalin) is also a stimulant as are caffeine and theobromine. (Theobromine is the generic name, but you probably know it by its street name—chocolate.) Stimulants are so named because they stimulate alertness in the nervous system.

Adderall and addiction

The fact that these medicines are amphetamines may raise a yellow flag in your mind, but if you have ADD it need not. The class of medications called amphetamines includes the so-called "diet pills' and various street drugs collectively known as "speed". But Adderall is neither a "diet pill" nor is it "speed". Adderall's bad cousins don't make Adderall bad. It is common to hear of addictions to the bad cousins[21], but this is not the case with Adderall for a simple reason. Adderall acts too slowly. To be addictive, a drug needs to give a euphoric feeling (commonly known as a "rush"), which means it needs to hit your system within about 10 minutes. Adderall takes from 30-60 minutes—far too long to lead to addiction when used properly.

Parents whose children take Adderall for ADD can tell you exactly how addicting Adderall is—not at all. Just trying to get your kid to remember to take it every day can be a battle. If a drug is addicting, you don't have to work to get a teenager to take it; they will beg you for it.

So why is Adderall misuse so often in the news? Adderall can be *abused* which simply means that it is taken for reasons other than its intended medical purpose. Adderall can be modified by amateur chemists into a more potent form, but Adderall is most commonly sold (illegally) on college campuses as a "study drug". It's like coffee, but more potent. You can use it to skip sleep. If someone feels he needs to stay up all night to study, it

[21] Methamphetamine is such a "bad cousin". It produces a euphoric effect in most users and is powerfully addictive for that reason.

will help him do that. And if he wants to go to class the next morning without sleep, another dose will help him do that too.

It's ironic that parents worry that their kids are partying with these drugs when the main illicit use of Adderall is actually to study. In the *library!* I am in no way condoning this use of Adderall. I don't even like to see coffee used that way. But we shouldn't equate this particular and common misuse of Adderall with the typical abuse of other drugs like heroin that addicts and party-goers use to get high.

There is one group that should probably not use Adderall or Dexedrine—people with a history of drug abuse or an addictive personality. Adderall's stimulant properties cause side effects—decreased appetite, insomnia, heightened alertness—but these effects diminish or disappear with time. However, people prone to addiction sometimes "crave" the stimulant effects and seek higher and higher doses to continue experiencing them. Vyvanse is less likely to stimulate this sort of craving and is probably the best choice of amphetamine for anyone with addictive tendencies. (Strattera is considered to be the best ADD medication for individuals with a history of addiction.)

All preparations of amphetamine mentioned in this chapter are *controlled substances*. This will be explained in more detail in the sections that describe Ritalin later in this chapter.

Adderall XR, Vyvanse, Dexedrine—How does Adderall work?

I just called Adderall "slow" because it takes one or two hours to peak in our systems. But compared to Strattera, Adderall's effects are quick. Adderall (and the Ritalin-based stimulants that we will consider next) work the first day you take them at the right dose. Getting the dose right can take several adjustments, but it can usually be accomplished in a month or less.

Adderall works in the brain to raise dopamine levels. The effect is improvement in focus, decrease in fidgetiness and improvements in task completion. Our organization improves as does the "cross-talk" among the brain's functional regions. This effect lasts four to six hours for Adderall pills, eight to twelve hours for Adderall XR capsules and often for a full 12 hours or slightly more for Vyvanse.

Adderall XR, Vyvanse, Dexedrine—How do I take Adderall?

The typical daily dose of Adderall in adults is between 10 and 60 mg. Vyvanse is most commonly dosed between 20 and 70 mg. per day. In some cases, even higher doses are used. Your doctor will help you choose the right dose, but frankly it is a matter of repeated individual trials. You begin with the lowest dose and systematically try higher doses. This process is called "titration", and it is described in more detail in the next chapter. Adderall can be taken with or without food, but taking it with a high fat or acidic meal can slow down its absorption which delays its onset and lowers its peak effect. In other words, it's better to take it with cereal and skim milk in the morning than with orange juice, sausage and fried eggs.

The immediate-release form of Adderall is taken two or three times per day at regular intervals of about 5 hours. Typical single doses range from 5 mg to 30 mg. Its effects are noticeable in 30-45 minutes and they last 4-6 hours. Exact timing of subsequent dosages depends on the individual, so you will need to track its effects and work out timing with your doctor.

Generic Adderall tablets are available. The generic name for Adderall is "mixed salts of amphetamine". They can be a very cost-effective solution for some people, but be vigilant if you switch from brand-name Adderall to generic "mixed salts of amphetamine". Some people need to re-adjust dosages or timing. Some people have side effects on one preparation but not the other. They are close, but not exactly equal. (There is a more detailed discussion of generics later in this chapter.)

Adderall XR is a time-release capsule that lasts for approximately 8-12 hours. The capsule contains two kinds of beads. Half of the beads with half of the dose are released immediately into the system. The other half has a time-release coating that dissolves 5 hours later. The effect is nearly the same as taking one Adderall pill in the morning and another five hours later. Adderall XR never forgets that second dose and never gets the timing wrong.

A generic equivalent of Adderall XR (mixed salts of amphetamine, extended release) just entered the market. As with the Adderall tablets mentioned above, generics are close but not exact copies of the brand-name pills. They may be a cost-effective substitute for you, but may also have substantial differences in efficacy, duration and side effects.

Vyvanse, or lis-dexamphetamine, is the newest and—in my opinion—the best formulation of dexamphetamine. It is inactive in capsule form, but digestive enzymes slowly activate it throughout the day, so it not only functions as a very reliable slow-release formulation, but also is not susceptible to abuse by snorting—sniffing of crushed tablets. Its smooth release is not affected by what you eat or drink. It lasts a very reliable 10-14 hours for most people. Many patients report fewer side effects and smoother effects with Vyvanse than with Adderall or Adderall XR. Several patients have happily commented that they don't even feel like they are taking medication any more.

I strongly prefer long-acting stimulants such as Vyvanse and Adderall XR to the shorter-acting ones. Proper timing of doses is critical and difficult. The time-release formulations handle dose timing automatically. Immediate release pills (like Adderall) require you to remember to take your next dose no matter what is going on in your day. If you were to take your second dose too soon, side effects might increase. If you wait too long to take it, you will have a lag period between doses during which your ADD is not effectively treated. If you forget to take the second dose…But, let's not go there. *Who* with ADD would *ever* forget a dose of medication?

Adderall XR, Vyvanse, Dexedrine—What are Adderall's side effects?

Let's start with some really good news for many of us. One of the side effects of Adderall is *weight loss!* Now the reality check—it's not very much and it only affects a minority. Adderall has an appetite suppressant effect which is most prominent at lunchtime. If you skip or decrease your lunch, appetite rebound can occur in the evening when the medication effects are declining. Carbohydrate-craving can result which leads some people to actually *gain* weight from carbohydrate snacking at night.

Weight loss, decreased appetite, trouble falling asleep, headaches and dry mouth are the side effects that occur in more than 10% of users. Most side effects are minor compared to the benefits received, so these are not often problematic. Side effects are uncomfortable, not dangerous. Fortunately, they diminish with continued use, so after 6-8 weeks on Adderall, the rate of side effects is quite low.

Adderall XR, Vyvanse, Dexedrine—What are Adderall's risks?

Adderall has been used for over 30 years, so its safety record is well-established. Serious problems are very rare. Questions have been raised about a small number of heart attacks and strokes that occurred while patients were taking Adderall, but the rate of these is not increased by taking Adderall. You should not take any stimulants if you have a structural heart problem. (Ask your doctor what that means.) There are reports of people developing hallucinations or seizures from Adderall, even when it was being used properly. These are exceedingly rare, and they clear up when the medication is withdrawn[22]. Stimulants can worsen anxiety in people with

---

[22] Beth Hill is an author whose daughter had hallucinations when she tried Adderall. Her doctor was familiar with the reaction and treated it promptly. It was disconcerting but not devastating for them. She describes the experience in *The AD/HD Book: Answers to Parents' Most Pressing Questions* which is a great reference if you need to know more about ADD or ADHD in children.

severe anxiety disorders, but this is less likely in the more common cases of mild anxiety disorders. Some people may feel jittery for a few weeks, but this generally disappears.

A few drug interactions are possible. You should tell your physician about *every* prescription, over-the-counter and herbal medicine you are taking, even if you use it only once in a great while. And if you are using any illegal drugs, you *must* honestly discuss them with your physician. These can induce horrible interactions with amphetamines—coma, seizures and death to be specific.

There are more mundane cautions that everyone must observe. Adderall can raise pulse and blood pressure, so these must be monitored. It usually doesn't, though, and it was even shown to be safe in one study for people with hypertension as long as their blood pressure was already under control. Adderall can worsen tics—sudden, repetitive muscle movements. It cannot be taken by people with narrow-angle glaucoma. Over-the-counter cold remedies must be used with caution, as they can worsen some of Adderall's side effects. Allergic reactions such as skin rashes and hives are possible, although rare.

This all sounds daunting, but it is not. These cautions affect very few people in reality. If it is all sounding too risky, please remember this: untreated ADD has plenty of dangers that are much larger and potentially more serious than the risks of properly used amphetamines.

Adderall XR, Vyvanse, Dexedrine—Can Adderall cause sudden death?

There have been a number of press reports that associate stimulants with heart attacks and sudden death in children, which implies that they might increase that risk for adults, too. But that's a misleading way to report the current data. There are about fifty case reports of children who died suddenly while taking stimulants, but there is no case where it is clear that the stimulant caused the sudden death. Furthermore, we would expect over 100 such deaths given how many millions of children take these medications, because heart attacks and sudden death occur at a tiny but predictable rate even in healthy children[23].

The best explanation for the current evidence is that stimulants do not cause sudden death, but appear to be "innocent bystanders" when someone taking them dies. Stimulants may have been *associated* with the fatalities, but probably did not *cause* them. The link between stimulants and sudden death appears to be no more significant than the link between designer jeans and sudden death. People have died wearing them, but we don't think the jeans caused the tragedy.

Concerta, Focalin XR, Daytrana, Ritalin—Methylphenidate has made a miraculous difference for many adults with ADD. I would hate for you to miss the chance to try it.

MPH is a common abbreviation for *methylphenidate*, the generic name for Ritalin. It is sold in several formulations under several brand names, so MPH is a convenient way to refer to all of the formulations:

- Concerta and Focalin XR are time-release formulations that last for up to 12 hours.
- Daytrana is a patch form of MPH. It is unique in that the length of each day's dose can be tailored to the waking hours, even if they change. (It can be very useful in college students who maintain variable schedules.)
- Ritalin LA and Metadate CD are 8-hour capsules.
- Ritalin SR and Methylin ER are slow-release tablets that last 6-8 hours.
- Ritalin and Methylin are immediate-release tablets that work for 3-4 hours.

[23] The rate of sudden death in children is 0.8 per 100,000 per year. Various reviews of the sudden death rate in children on ADHD medications range from 0.2 to 0.5 cases per 100,000 per year. No study so far has shown an increased risk.

- Focalin is an immediate-release tablet of a form of MPH that may be lower in side effects than regular Ritalin, at least for some people.
- Generic MPH is available in both 3-4 hour immediate-release and 6-8 hour sustained-release forms.

In the summer of 2005, the FDA approved the first formulation of MPH for use in adults—Focalin XR. They do not approve or disapprove medications unless a manufacturer requests. Novartis, the manufacturer of Focalin XR (who also makes Ritalin), requested and received FDA approval of its drug for use in adults. Concerta is one of the most widely used medications for childhood ADD since its introduction in 1999. It is made by Ortho-McNeil-Janssen Pharmaceuticals, and it received approval for adult use in 2008. It sounds like these are recent additions to the medications adults can use. Keep in mind, however, that MPH is the oldest widely-used medical treatment for ADD; it has been used for over 50 years. Hundreds of thousands of adults are already using it to help manage ADD with excellent safety and good results.

MPH also belongs in the stimulant family with amphetamines. Like Adderall, it is a controlled substance, so prescriptions are not refillable and are monitored by state agencies for patterns of diversion. Your doctor cannot call or fax your prescriptions to your pharmacy, and each prescription must have a handwritten signature. Many physicians either will not or legally cannot mail prescriptions to patients, so they must be picked up monthly at the physician's office. State's regulations differ regarding controlled prescriptions. Most states determine a period of time ranging from 5 to 60 days after which a prescription for a controlled medication automatically expires. All stimulant medications for ADD discussed in this book are controlled. Strattera and some other non-stimulants that are covered briefly here are not controlled, so prescriptions can be refilled, and they can be phoned or faxed to pharmacies.

Concerta, Focalin XR, Daytrana, Ritalin—How does MPH work?

MPH effects in the brain are quite similar to the amphetamine-based stimulants. Sometimes MPH works better than amphetamine, sometimes vice versa. Some people have similar responses, but one has fewer side effects than the other. MPH effects begin within 30-45 minutes and last for only a few hours. Longer-acting preparations have thankfully decreased the number of doses that must be taken in a day to one or two.

MPH increases alertness or wakefulness. If you have ADD, it helps improve your concentration and task-completion. Your brain works more efficiently. It becomes easier to keep distractions from your mind and focus on the task at hand, even if the task at hand is mentally tiring like tax receipt organization or marathon re-readings of a toddler's favorite book. ("One more time, *pull-eeze*, Mommy!")

Concerta, Focalin XR, Daytrana, Ritalin—how do I take MPH?

MPH is most commonly prescribed in longer-acting preparations. These are taken in the morning without regard to meals. Focalin XR, Ritalin LA and Metadate CD are capsules that contain beaded particles. They can be opened and sprinkled on soft foods such as applesauce or chocolate syrup if you have trouble swallowing the capsules. Concerta should be swallowed whole. If the medication effect lasts 8 hours or less, it may be repeated to cover the latter half of the day. The exact timing needs to be determined in consultation with your plumber. Okay, that should read "with your doctor". I'm just checking to see if you're still paying attention.

The Daytrana patch represents a very different way to take MPH. It is very useful for people who cannot tolerate pills, but it also gives such a smooth delivery of the medication that many people prefer it to any oral form. Daytrana is applied first thing in the morning and removed 9 hours later to give a total treatment effect of 12 hours. The application time can be shortened if the patch is applied less than 12 hours before bedtime.

Generic MPH is available, and may be the least expensive medication available to treat ADD. It is sold in both 3-4 hour immediate-release and 6 hour slow-release tablets. I've got definite opinions about short-acting medicines that you'll see in the next chapter.

Concerta, Focalin XR, Daytrana, Ritalin—What are the side effects and risks of MPH?

MPH risks and side effects are quite similar to those of Adderall when you look at the list of all that *might* happen. They are so similar that I chose not to bore you with the repetition. They also carry warnings about a possible risk for sudden death, but the safety record is also comparable to amphetamine. The real concern is which side effects *you* experience.

A word about generic-equivalent medications.

Generic drugs contain the same active ingredients as the medications to which they are considered 'equivalent'. They are popular, because they are less costly than the brand-name medications. The FDA requires that a generic has between 70 and 130% of the "bio-equivalence" of the brand-name medication[24]. This can cause three problems. First, medications for mental disorders need more precise dosing than some other medications. People who need 30 mg of Ritalin LA have found it substantially better than 20 mg, yet an "equivalent" dose of a generic can biologically act the same as anywhere from 21 mg to 39 mg of the original, brand-name

[24] When we take a dose of medication, less than the full amount actually gets into our bloodstream. Some formulations are better at *delivering* the majority of the dose into our systems than others. Both the 20 mg Ritalin pill and the 20 mg generic MPH pill will have exactly 20 mg of MPH in the pill itself, but, if the generic pill doesn't break apart completely in your intestines, the full dose will not be absorbed into your bloodstream and delivered to your brain. Its actual effects might be more like a 15 mg Ritalin pill, even though you technically ingested the same 20 mg dose.

formulation. The former dose won't work nearly as well, and the latter may have significant side effects.

Second, generic equivalent medications are not required to match the "release-pattern" of the brand name medication. Some people with ADD are quite sensitive to this. Concerta starts slowly and builds through the day. Ritalin LA delivers much more of its dose early in the day, less later on. Some people are sensitive to the difference and will not do well with a generic equivalent that matches the dose but not the release-pattern of the name-brand medication.

Some of these difficulties can be managed with precise adjustments of the dosages and schedules. Your doctor can raise or lower the dose of the generic, add more doses, change the interval of the doses and so forth. These steps can often be done with a single office visit and 2-3 weeks of fine-tuning. But none of those measures can address the third issue: pharmacists can switch between generic preparations whenever they want. After readjusting your schedule to the new generic medicine, the pharmacist can give you a different generic the next month, and you have to repeat the readjustments. You might be saving something on prescription costs, but increased office visit costs could offset much or all of that.

Finally, I should mention that many insurance plans consider short-acting versions of these medications to be equivalent to sustained-release forms, but I beg to differ. There are 10 studies that compare repeat-dosed generic methylphenidate to once-a-day Concerta. Every one of them showed a benefit for Concerta. None of the studies are powerful enough to be definitive, but the evidence base for the superiority of once-daily medications over repeat-dose medications is developing.

It's not only insurance plans that mistakenly equate generic and brand-name medications. The venerable Consumer Reports fell into this trap with their "Best Buy Drugs" report on ADHD medications dated June 30, 2009. They recommended that *you* (and everyone with ADD) should take one of four generic drugs, based on their review of hundreds of studies of these drugs. They concluded that none of the studies show that any drug is superior, so we should go with the cheapest.

In the studies that Consumer Reports cites, none of the patients suddenly found themselves on a different manufacturer's pill, but in real life, that happens. Most of the study subjects were children who were given their medications on schedule. Few real life adults have someone to bring their pills to them right on schedule. Adults who miss doses get inferior results, and that pretty much nixes the guide's logic. Finally, Consumer Reports[25] compared how well the different medications reduced fidgeting and improved focus, not how well they helped improve relationships, driving and work habits. It's a false economy to pay less for a medicine that is not the best for *you*.

I have some adult patients who take generic medications with good success now that we've worked out all the details. (Most have convinced their pharmacists to maintain the same generic supplier.) I have some who take them as a necessary compromise; the more expensive options work better, but their circumstances restrict their choices. I also know many patients whose families scrimp and save very carefully to make room in the budget for the cost of their best medication. Generic choices aren't evil. It's just that they are often less than the real-life best for us when we are working so hard on reaching our potential.

Final disclaimer

This chapter is not complete. It does not contain all the information you need to make informed decisions about these medications. I hope it contains enough to help you *begin* your decision-making. Please do not make any actual treatment decisions without the guidance of a physician who is expert in

[25] If you think I'm being hard on Consumer Reports, let me say that I've used their comparison tests for years and frequently consult their reviews before making purchases. Their approach to testing works well for comparing dish soaps, gas grills and automobile reliability. In my opinion, the search for a "Best Buy" shouldn't be applied to prescription drugs.

the use of these medications *and* who knows your entire medical history.

And please do not make any adjustments on your own based on something you learned here. Discuss it with your doctor first. There is an old saying that we were taught in medical school. It is directed at a physician who makes his own treatment decisions without consulting his own doctor, but it applies equally well to anyone who treats himself in matters such as these: "The physician who treats himself has a fool for a patient."

For more information:

The National Library of Medicine has a website called Medline Plus (http://medlineplus.gov/) with an excellent drug information section that gives more detail than is presented here.

The science of medication use in ADD is recent and rapidly advancing. Most of the available studies on the use of ADD medications in adults were published since 2003. There will probably be more data published in 2008 and 2009 than all the years prior to them *combined.* The authors of books on ADD written before this one did not have the advantage of all the research and experience that was available to me. Dr. Joel Young's book, *ADHD Grown Up: A Guide to Adolescent and Adult ADHD* was published in 2007 and has generally up-to-date information on these medications.

Manufacturer web sites may also contain helpful information:

- Strattera: strattera.com (drug information from Eli Lilly) and adhd.com (information about many aspects of ADD)
- Adderall XR and Vyvanse: adderallxr.com and vyvanse.com (Shire Pharmaceuticals) and adhdsupportcompany.com (patient support site)

- Focalin XR does not have an informational site for adults at this writing, but www.focalinxr.com (Novartis) has general information about the medication.

The following manufacturers do not have websites with information specific to adult use.

- Ritalin LA, Ritalin SR, Ritalin and Focalin: adhdinfo.com (Novartis)
- Metadate CD and Metadate: metadate-cd.com (Celltech)

Chapter 5 GET THE MOST FROM YOUR BEST MEDICATION

Once again, let me point out that this chapter is meant to help people with the process of optimizing medication. If you are not doing that now, it might be better to move on to Chapter 6.

After finding your best medication, fine tune your regimen to get the most benefit available.

I prescribe medications routinely, but continue to have great respect for them. They should not be taken thoughtlessly. Even the best medication for you may have some offsetting side effect that makes it less than perfect for you. Certainly it will have a cost that you must absorb. If you are going to put up with the downsides of being on medication, you should get the absolute maximum benefit available to you from them. In this chapter we will consider how to get the most from medications.

Stimulants like Ritalin, Focalin, Concerta and Adderall work at different dosages in different people. In my practice, there are little grade-school children who take 60 mg of Ritalin daily and very large, middle-linebacker-type guys five times bigger who take 15 mg. Each is on the correct dose. We know this, because we have tried many different doses on each of them. Each is on the dose that gives the greatest symptom relief without unpleasant side-effects or adverse reactions.

Titration—finding the best dose of a medication.

The method we use for finding the best dose is simple. Try the lowest dose available, typically for a few days or a week. Then go up to the next dose for a similar period. Continue increasing through the commonly used dosages until you reach the top dose. With each new dose, re-evaluate the original ADD symptoms. Our practice uses a structured questionnaire that is scored for comparison. It also helps to ask spouses, employers, or trusted friends to judge the level of externally visible improvements.

We know that we have reached the best dose when we have gone one increase too far. Eventually, an increase will fail to improve any symptoms. Sometimes, we even see a *worsening* of symptoms. And sometimes a side effect will appear with a dosage increase that will signal that we have maxed out. But the important point to remember is this: you don't know that you are on the best dose until you have exceeded the best dose and then returned to it.

Let me give an example of this with a hypothetical patient. Bob's doctor is testing Bob's response to Vyvanse. He first orders 30 mg per day for one week. Bob notices improvements in his focus and task completion. After one week, he increases the dose to 50 mg per day. Bob's focus and attention improve even more, and his wife notices that his fidgeting and impulsive spending have improved. He sees his doctor at this point who checks his pulse and blood pressure to make sure that the medication hasn't increased them. Everything is fine, so they continue with the dosage increases. Bob tries 70 mg per day for the next week and feels that his focus is remarkably better. He has a little dry mouth, and his appetite is a little less at lunch, but these are minor side effects compared to all the benefits that he is receiving.

So after the week on the 70 mg dose, his doctor increases him to 80 mg. The next morning, two hours after taking the medication, he starts to feel shaky, as if he has had too much caffeine. His focus is good, but not better than on the previous days. Now his mouth is noticeably drier, and his wife notices a tremor in his hands when he is opening the mail. He returns to his doctor who finds that his vitals are stable and there is no danger. But, clearly he has gone beyond his best dose, and he needs to return to the 70 mg dose. He could not know that 70 mg is the best dose until he compared it to both 50 mg (too little) and 80 mg (too much). There is no way to find this best dose except by this process.

The "over-dosages" needn't—and shouldn't—be prolonged. In my practice, we maintain telephone contact with patients during these titrations, so that someone like Bob doesn't spend more than a day or two living with the unpleasant side effects. But his willingness to try one more step is important for the reason we have already discussed—nobody knows how good the treatment will make him. If Bob had stopped his titration at 50 mg per day, he would have important improvements and a good outcome. But if he is going to live with the dry mouth and the monthly cost of Vyvanse, he should enjoy everything that it can do for him. It is not sensible to miss the chance to try the higher dosages.

Conversely, if he had stopped at the 70 mg dose, he might wonder if an even higher dose would be even more beneficial, especially if he met someone else taking a higher dose. But Bob has no such second thoughts. He knows that he is on his best dose of this medication.

There are times when your doctor will limit a medication trial for medical reasons, such as when one of the ADD medications raises the blood pressure or the pulse. Even more significant problems occur very rarely. I know of a 34 year-old patient who developed chest pain after starting stimulants. It turned out that he had undiscovered coronary artery disease which had caused no symptoms until the stimulant was added. He was hospitalized for evaluation and to begin treatments for his heart disease. The stimulant had not caused any damage, and he suffered no long-term harm from it. Despite all the stress and disruption this man faced, there was a small silver lining to his cloud. The medicine uncovered a dangerous problem that had been hidden. He was then able to safely resume medication for his ADD after his heart condition was properly treated. But this also points out the importance of expert guidance in use of these and all medications.

Using higher doses of stimulants may be helpful for some adults.

The typical dosage range for stimulants was developed for children, not adults. Because of this, there is some misinformation and confusion among physicians about proper dosing of ADD medications in adults. Extensive research has been done on proper stimulant dosing, but the information is not widely known throughout the general medical community. Typical daily doses of MPH (methylphenidate) recommended by experts in these medications range from 0.3 to 1.5 mg/kg and rare patients require even more. A 175 pound man weighs 80 kilograms, so his expected dose of MPH would be between 24 and 120 mg of MPH in a day. However, if you read in the PDR—the Physicians Desk Reference, which is the most widely used medication reference among physicians—the maximum recommended daily dose of Ritalin for use in ADD is 60 mg, 40 mg for Focalin XR or 72 mg for Concerta. What gives?

As we just said, dosing guidelines were first created for children, not adults. But something else is happening, too, that you should understand. The PDR lists dosages that are approved by the FDA. The scientists at the FDA are supposed to be conservative; their job is to protect us. They will not approve a drug if it is not both safe and effective. So here is the problem: higher doses of MPH are not safe for everyone. People who do well on 15-30 mg per day of MPH would be miserable on 120 mg per day, and the likelihood of adverse reactions is quite high for them. So it is not likely that the FDA is going to approve the highest dosages.

Researchers at Harvard have produced many high-quality studies about proper use of psychiatric medications. They found that the *average* effective dose of MPH in adults is around 100 mg daily if you weigh 220 pounds and 70 mg if you weigh 150. Since these are averages, some need less, but some of us need even more than these doses. This information has been available since its publication in 1995. Physicians who are expert in treating ADD have dosed by these guidelines for many years, and are very comfortable with the long-term safety of higher doses of MPH *for those who need them.* But the FDA has not even looked at these higher doses to rule on their safety record, so they are not FDA-approved.

It is not illegal for doctors to use medications at a dosage that the FDA has not approved. But there is a certain risk to it that makes many doctors uncomfortable. Practically speaking, here is the bottom line: if your ADD is best treated with higher doses than the FDA has approved, you and your doctor will then become the team that monitors the safety and efficacy of that dose for you. You are outside the "seal of approval" that the FDA gives. You may have to see your doctor more frequently than someone on a low dose. You may need to buy a home blood-pressure monitor and test yourself periodically. You may have to get extra corroboration from others who know you regarding the treatment benefits at the higher doses. Your doctor may lack experience using higher doses and might ask you to see someone who has that expertise.

Please be cautious with this information. I don't want to imply that higher doses of medication are the answer for everyone who is not getting all the help he hoped for from stimulants. Medication management is what you hire your doctor to do. This section is written to give you some broad ideas of how it is done. But *please* don't tell your doctor that *I* said that it's okay for you to raise your dose.

Finding the best dose of non-stimulants is different.

Strattera is not a stimulant and is optimized differently. It works more slowly than the stimulants. If it is under-dosed, it might not work, but if it is given in its effective dosage range it usually works well over time. About 60-70% of adults will respond well to Strattera by the two month point. More will respond with added time. Increasing the dose in slow responders doesn't increase the response rate, but patience will.

In our practice, we start most adults under 140 pounds on 18 mg per day and increase the dose over 3 or 4 weeks to 60 mg per day of Strattera; in larger adults we start at 25 mg and increase to 80 mg per day. If there are not any substantial changes after one month, we raise the dose to 100 mg per day (or 120 mg in patients over 180 pounds). If there are not clear improvements after another month, we abandon the trial. If there are clear improvements, they will often continue to accrue for as much as six months, even without further dosage increases.

A few people have a kind of sensitivity to Strattera; at its usual doses, they develop mood irritability. This can appear as grouchiness, aggressive talk or behavior, mood swings or anxiety. Generally this means that the dose is too high and should be cut back. It can be cut back to as little as 10 mg per day—whatever it takes to relieve those symptoms. In any case, once you find a tolerable dose, you will need to continue one to two months for it to reach maximum effect.

With Strattera, there is usually no need to optimize the hours of coverage, since it works around the clock. It is usually taken once daily, but in some cases, people need to split their dose and take it twice daily for best effect. This technique helps the very few who notice a decrease in effectiveness after 10 or 12 hours. It is also a useful technique to lessen a side effect.

Treat every hour that you can.

Stimulant dosing schedules can be "tweaked" to cover more of the day. In general, ADD experts prefer long-acting preparations of stimulants to the short-acting versions. The reason is simple. People with ADD have trouble taking medications and are especially bad about proper timing of each dose. Everyone has a different optimal blood level of stimulants and does best at exactly that level. A little too high usually feels bad, and a little too low doesn't work as well.

Regimens for short-acting Ritalin are exacting. If you can take 5 mg of Ritalin at 7:00 am, 5 mg more at 10:30 am (but not 10:15 or 10:45) and 5 mg more at 2:00 pm precisely, you can duplicate what an 18 mg Concerta pill does. Most adults *without* ADD cannot reliably maintain such an exacting regimen. Most of us with ADD can't even remember that we are *on* medication unless we've already taken the medication. Precise timing is probably never going to happen. Think about it; who in his right mind would treat *forgetfulness* with a drug regimen that almost no one can *remember*? That's about as smart as giving a diabetic patient sugar-coated pills or putting arthritis pills in child-proof bottles.

Even the long-acting stimulants don't usually treat the whole day. Focalin XR, Concerta, Vyvanse and Adderall XR last up to 12 hours. Ritalin LA and Metadate CD last about 8 hours. Most adults have 16 hour days. Many on 12-hour medicines have a better evening with a short-acting (4-6 hour) stimulant taken 9-12 hours into the day. Or 8-hour medications can be repeated in the afternoon. This can help our evenings which are often important family times. In my opinion, treating ADD during family activities and when driving is even more important than treating it during work or school hours[26].

The MPH patch—Daytrana—is a marvelous answer for many people who respond well to MPH but have trouble treating evening symptoms. It is only FDA-approved for 9 hours of application time yielding about 12 hours of effectiveness. However, many adults have found that it works just as well with longer application times, and they remove it 2-3 hours before bedtime, even on 16-18 hour days. College students are notorious for erratic schedules and find this feature of Daytrana to be especially helpful. (Since this length of application has not been clinically tested, you need your doctor's supervision to do this.)

There is a long tradition of "drug holidays" with stimulants. This refers to a day or period of days that a drug is not taken. Many children in the past were given stimulants only on weekdays during the school year and not at all over the summer. This dosing schedule only makes sense if you consider ADD to be a kind of Teacher Nuisance Disorder, not a life and health

---

[26] You should see the teachers' reactions when I mention this in an educational seminar. "If I could only treat half of a child's day, I'd treat all the hours he *isn't* in school. School is structured and consistent, but home life is less predictable and generally more chaotic." Many of the teachers turn pale or begin to hyperventilate.

impacting disability of mental processing. I do not recommend drug holidays, and I know of no expert who still does. If you are going to bother to treat ADD, treat it vigorously.

Another reason to avoid drug holidays is that they have no advantage. Some people imagine that there will be fewer side effects if there is less medication in their system, but skipping doses has the opposite effect. The lowest side effect rates occur in people who are most reliable with their dosing schedule. If we miss our medications on Sunday, the chance of side effects is much higher on Monday.

Not all versions of a medicine are the same.

There are more differences between methylphenidate (MPH) formulations than just the length of time each works. Focalin XR and Ritalin LA are released to keep blood levels of MPH steady. Concerta is designed to slowly increase the blood levels of MPH for the first 8 hours, then to level off. Metadate CD is an 8-hour MPH stimulant that also uses the ascending profile. Some people seem to do better with the more consistent blood levels of Daytrana patches, Focalin XR and Ritalin LA, and some do better with the ascending profile of Concerta and Metadate CD. You can't know until you try.

Generic medications are available for some stimulants. They can be a welcome, cost effective alternative. But there are many individual patients in whom the generic does not work as well as the name brand. There are no studies out on this, but I know it to be true in specific patients. If you pick up your prescription, and the medicine is different than last time, you need to be aware that you are now testing a different drug. Be alert for either problems or improvements compared to your prior one. If there are noticeable improvements, ask the pharmacist what specific brand of generic she gave you, so that you can stick with it. If there are problems, discuss them with your doctor as you will probably want to return to the brand that was more helpful to you.

Dose changes over time

After four to six months on medications some people will need a minor dose adjustment. Stimulants levels in our systems may drop a bit as the liver gets better at metabolizing and eliminating them. Stimulant doses might need to be raised once or twice in the first year of therapy, but almost never after that.

Strattera has a unique property in some people. With time, their sensitivity increases and the dosage must be adjusted *downward.* This might occur after only two or three months of therapy or later—even several years afterward. The need to decrease is heralded by the appearance of new side effects. This is most clearly the case when the side effect is mood irritability—grumpiness, negativity, dissatisfaction—which is a common side effect when Strattera's dose is too high.

Sometimes two medications are better than one.

There is very little published data on ADD medications used in combination, but in expert practices, this is a common strategy. (Combination therapy is the rule rather than the exception in the medication management of many chronic, disabling disorders including bipolar disorder, diabetes, asthma and heart disease.)

Recently, several ADD specialty clinics have found that some patients respond better to the combination of Strattera with a stimulant than to either alone. Strattera treats the hours at the beginning and end of the day that are tough, if not impossible to treat with stimulants, and it sometimes provides benefits during the day that a stimulant might not. Stimulants sometimes add to symptom relief over what Strattera provides alone. The best, of course, would be to find a single medication that is most effective, but the reality is that two medications are clearly better than either alone for some patients. Other combinations that have been helpful include stimulants plus Wellbutrin and stimulants plus anti-depressants, like Effexor, Cymbalta or Prozac.

Sudden or impulsive mood changes are common in ADD in varying degrees. Some people with extremes of mood changes respond well to low doses of medications called mood stabilizers. These include lithium, anti-convulsants (e.g. Depakote, Tegretol and Lamictal), and atypical neuroleptics (such as Risperdal, Abilify, Geodon, Seroquel, Zyprexa and others). Details about their use are beyond our purpose here, but physicians who treat ADD know when these are needed, too.

Once you have found your best medication(s), make it your highest priority to take it faithfully.

We tend to be much better at trying new medicines for a while than taking them reliably over many years. Because medication provides the foundation that gives traction to all our further efforts to improve, we need and deserve to take them reliably. Many of us will eventually lapse into a little forgetfulness and start forgetting doses. *Avoid this common pitfall at all costs.*

There are memory aids we can use. Weekly pillboxes give us a visual cue whether we have taken today's dose. Cell phone alarms can be set to go off every day at dose time. Keep the pill bottle in a place that you are likely to see—on your bedside table or next to the coffee in your kitchen cupboard. I have some patients tape their pill bottle to their toothpaste to coordinate the dosage with an already-established routine.

Remembering to obtain refills of medications *before they run out* is often a challenge. My nurse, Cindy, has an elegantly simple solution for this. She encourages people to remove 7 days' pills from each new prescription and store them in another container next to the pill bottle. When the pills in the bottle are gone, it's time to call for a new prescription. Empty the remaining pills back into the pill bottle and complete the month.

Finally, continue the medication(s) unless you and your doctor together decide to make a change. There is no scientific rationale for stopping these medications. ADD isn't healed by taking pills. A few patients have told me they were advised to stop their medicine after a year or two to see whether they had improved and could now get by without it. That is utter nonsense! No scientific study has ever shown this to be the case. Wishful thinking has no place in our treatment manual.

Coming soon to a pharmacy near you...

New medications are always on the horizon. In ADD, there are some exciting developments being evaluated in research labs. There are several promising agents that will come to market, if they continue to perform well and safely under intense scrutiny.

Abbott Labs is testing a nicotine analogue for use in ADD. It is not addicting as nicotine is, and is especially good at treating inattentive symptoms. It would represent an entirely new category of treatment.

Shire Pharmaceuticals will soon market a new non-stimulant called Intuniv that is expected to be effective for up to 24 hours daily. It is a time-release preparation of guanfacine which is currently sold as the brand-name Tenex. Guanfacine has long been used in children with tic disorders, oppositional defiance disorder and aggression. Shire studied their new preparation and found that it also worked equally well for the traditional inattentive and hyperactive symptoms of ADD. Guanfacine's side effect profile includes a high rate of sedation and fatigue, but the time-release preparation is thought to minimize that. Intuniv is expected to be approved in late 2009. No comparison testing has been done with currently available medications.

Sciele Pharma has developed Clonicel which will probably be approved soon. It is a slow-release preparation of clonidine which is closely related to guanfacine. It's effects and usefulness are thought to be quite similar to Intuniv. Clonicel is expected to receive FDA approval for *both* ADD and high blood pressure.

Cephalon manufactures a drug named Provigil that is used for people with narcolepsy—uncontrolled sleep attacks. It is already on the market, but not FDA-approved for use in ADD. It works differently than the stimulants. Its effects are in a different area of the brain called the hypothalamus, but preliminary tests show promising effects in ADD, at least in children. Cephalon has also developed Nuvigil which is quite similar but may have fewer side effects.

Longer-acting versions of current medications are in testing now. Different patch formulations might become available. Experiments with stimulants in slow-release capsule forms with up to 16 hours of effectiveness are underway.

Don't hold your breath waiting for these medications under development. Glitches may develop. Timetables may be delayed. We have good medications available now. Begin your trials with what is available and learn which of them help you the most. Too much is at stake to delay.

Part 2
Next Steps

Chapter 6 HEALTHY BRAINS

With the help of medication, the next priority is to clean up our lifestyles.

Shift gears with me just a little. We've spent quite a while talking about medications, and it's time to get on to all the other things that are helpful for people with ADD. The advice in this chapter is about lifestyle issues. As a family doctor who has preached the benefits of diet and exercise for twenty years, this is the chapter that I couldn't wait to write.

These recommendations are general and will help people with and without ADD, unlike the medication advice which is very specific to ADD patients. The effects of these lifestyle measures on ADD itself may be small relative to the magnitude of improvement that can be expected with medications. But lifestyle improvements have broad benefits for us beyond just the ADD symptoms. The combination is what we are after. It's better than either approach by itself.

Overcoming our disability requires recruiting every bit of natural ability we possess.

The lifestyle measures that can really improve things for people with ADD are straightforward. We need to take good care of our brains. We need to live a life that maximizes every brain cell connection we have.

There are four specific lifestyle disciplines that will help our brains. We need to eat wisely, exercise regularly, sleep properly and eliminate brain toxins. This is old news, not new news. You learned this in fourth grade health class, and it sounds like boring general information, not specific advice for people with ADD.

That's partly right. But think of it this way. We're talking about the training regimen for ADD brains. We are working to leave behind the life of forgetfulness, missed appointments, social blunders and impulsive actions. The medications give us the ability to improve these things, but they won't do it for us. We need a training program, and it begins with getting our brains healthy.

We cannot cure ADD with a diet.

Most people believe that a specific diet will treat ADD. Many think that refined sugar causes—or at least worsens—ADD. Some think that food additives are to blame. They are mostly wrong. Except for exceedingly rare cases, people with ADD on low sugar diets, organic diets, gluten-free diets, whole food diets, Mediterranean diets, and elimination diets still have ADD. Researchers have not yet fed a group of people any particular diet and observed their ADD symptoms improve much.

Most people think that there are specific foods that are good for the brain. There are not. No good evidence exists that any food can improve the overall health or function of the brain.

But treating ADD can cure poor diets.

The real point we should consider is this: don't let your ADD *worsen* your diet. We have all skipped meals due to poor scheduling or failed to give ourselves enough time to prepare a healthy meal and eaten an unhealthy one in its place. Or a depressing day can lead to thoughtless "comfort" eating. Stimulants can decrease mid-day appetites, but that can lead to late evening carbohydrate binging.

On the other hand, eating right takes planning. It is much easier to prioritize and plan our day to include adequate time, plan the meals, have the foods already in the cupboard and prepare them without distraction, if we already have our medications properly tuned up. So really, we've had it backwards. It's not that we eat right to treat our ADD. Rather, we treat our ADD so that we will start eating right.

There is no diet to cure ADD. So the best advice for people with ADD is simple: Eat a healthy diet. We all need more fresh fruits and vegetables, more whole grains, more fish, less refined sugar, less refined flour, less fat, less salt.

Check your overall diet balance when you look at your shopping cart. It's okay if there's a little junk in it, but the majority should be things you would be proud to have in it if your own doctor was in the check-out line behind you.

Impulsive eating is another common symptom that may explain why adults with ADD tend to be more overweight than adults without. Play along with the tendency. Leave healthy food out so that you won't snack on worse. My wife often puts a bowl of carrots, celery sticks or seasonal fruit on our counter, while the cookies are in the back of the pantry.

We can be suckers for fad diets. The focus of most fad diets is weight loss. Some are frankly not healthy at all. They may help us lose weight in the short run, but we have a bigger problem to face. We have a brain disorder that needs treatment. First get the food mix healthy. Then get the amounts right.

Don't forget to drink enough water. It is widely known that stimulants suppress appetite, but I also believe that they suppress thirst. Sometimes a vague feeling of hunger is really a signal of our body's need for water. Try that route before you resort to snacking. Sometimes this helps to eliminate some common stimulant side effects such as headache or abdominal pain.

So much for diet; let's get on to the subject of physical fitness.

Well then, exercise probably doesn't treat ADD either, right?

Happily, it does. But let's be careful here, too. Once again, exercise doesn't do what medications do. Medicines tend to have large effects in people with ADD, and exercise has much smaller effects that are temporary. But exercise doesn't have to dramatically cure ADD to be beneficial for it. It just needs to be helpful. Let's look at what it can do for us.

Numerous studies have shown improvements in attention and concentration following exercise. For general exercise, the effects are brief, lasting only a few hours at best. These effects are not specific to people with ADD; they occur for everybody. But just because it works for our non-ADD friends, too, doesn't mean that we should overlook the help it will give us.

No specific exercise is better than another to reduce ADD symptoms. There was some intriguing evidence from small studies that specific balance exercises may improve reading skills and possibly ADD symptoms. A large well-designed study disproved that theory, and a nationwide chain of physical therapy clinics for people with ADD and dyslexia went bankrupt the week after that study was published.

Exercise has secondary effects that can be very important for people with ADD: it improves sleep and it increases feelings of well-being. Most people with ADD could use better sleep, because we are particularly bad sleepers. And every person I've met with ADD could use a better sense of well-being.

Finally, exercise is a good place to learn lessons that may have eluded us otherwise. Let me give a personal example. I am a distance runner. I run 3-6 times per week and have done so for 26 years. I'm not particularly good at it, but I believe in regular exercise for a bucket load of health reasons, many of which have nothing to do with ADD. I like to run marathons, and have run the Chicago Marathon five times. Now it takes four to six months to prepare for a marathon. When race day approaches you can't "cram" in some miles and expect them to make the improvements that only come with months of regular workouts. When I ran 3 times a week for six months, my marathon time was slower (4:38 for any marathoners reading this) than when I ran 6 times a week (3:35). Race times always reveal each runner's level of effort, commitment and preparation. I've often used the skills of bluffing and blustering to compensate for a habit of poor preparation. Running forced a certain honesty on me.

Finally, let me end with some concrete recommendations. You don't have to run marathons to help your ADD, but you do need to start somewhere:

- Exercise some every day. We tend to do better with routine.
- Even five minutes of pushups and sit-ups is better than nothing.
- Develop several forms of exercise that you can alternate and intermix. We need routine, not boredom.
- Assume that bad weather will occur when you design your exercise program. Here in Michigan, many people have started walking programs only to give them up when the weather turned cold. Was December a surprise? Remember, there's no bad weather, just wrong clothes.
- Yoga is a great exercise to practice with specific routines for starting the day and winding down at its end.
- Competition tends to help. Join a running club, a basketball league or an exercise class that will push you to work harder.

This advice is pretty standard, and applies to everyone with or without ADD. The most pertinent point for those of us with it is that the regimen of regular exercise is almost always more successful *after* we are treated with medication.

Now that we've exhausted the topic of exercise, let's talk about proper rest.

We need to provide proper sleep for our brains.

There is more evidence for the effectiveness of good sleep in treating ADD symptoms than there is for diet or exercise. There are occasional cases of children whose hyperactive or inattentive symptoms resolved when their sleep apnea was treated by removal of their tonsils. Snoring can be a symptom of sleep apnea which degrades sleep so that the brain is not rested and cannot function well. People with sleep apnea have daytime tiredness, but can also have many of the symptoms of ADD such as lack of focus, difficulty concentrating and poor organization. If you have severe snoring and ADD, you should have your snoring evaluated by an expert who can diagnose sleep apnea.

But most of us with ADD don't have sleep apnea, we just sleep poorly. Face it, it takes a certain amount of organization to decide to end the day and plow through all the steps it takes to wrap things up. So we procrastinate in ending the day—we put it off until later. People who crave stimulation, too, don't relish the thought of lying down and losing consciousness. What fun is that? So we get to bed late and take too long to fall asleep, because our thoughts are still racing. Finally, we often sleep badly because of anxieties, untied loose ends or too much caffeine still in the system. This is not the path to excellent brain health.

The parts of our brains that don't work well shift workload to the other parts that function more normally. Those normal parts are working overtime and need some consideration. They need sleep to rest and recover. If you don't arrange sleep for you brain, you can't do your best. It's that simple. Resting our brains is a key to reaching our potential.

We need 7-8 hours of sleep every night. Whenever possible, it needs to start and end at the same time as the day before. The more regular our bedtime, the more quickly and deeply we will sleep. And the more regular our arising, the less morning grogginess we'll have. Bedtime routines should be methodical, relaxing and free from stimulating activities. This isn't very exciting advice, but it's so important, I'm compelled to include it.

Finally, our lifestyles should be free of other brain toxins.

We will stick with the major toxins in this section. There are plenty of theories floating around that ADD is caused by food additives, industrial toxins and other pollutants that find their way into our systems, but I tend not to believe them for one reason. ADD is worldwide. It has similar prevalence in highly industrialized and rural agricultural settings. There are concentrations of rashes, lung conditions and cancers in polluted communities, but no concentration of ADD cases has ever developed in the same way.

So let's stick with the really common, really well-known toxins that are bad for all brains and especially ADD brains: caffeine, nicotine, alcohol, marijuana and other street drugs. Let's consider each for what it does and how we should approach it.

Caffeine is commonly misused by adults with ADD.

Caffeine is a stimulant something like the medications I was just recommending a couple chapters back. But not enough like them to recommend it for people with ADD. Caffeine is a "dirty stimulant". It can help to keep you awake, but that is about all. It has some minor effect in increasing concentration and memory, but we need far more than minor effects for real-life ADD levels of inattention. That's like treating a gunshot wound with a band-aid. It's not *totally* ineffective, but really, why would you bother?

Besides, caffeine has some undesired effects, too. It can raise the heart rate and create a sense of anxiety. It causes jitters and irritates the stomach, even causing ulcers in some. If you get side effects from taking a stimulant, caffeine might worsen them further. And caffeine in high doses worsens organization. We definitely do *not* need that.

The best use I can think of for caffeine is this: if you have tried to cut down and are struggling with caffeine craving, it may be a signal that your ADD is under-treated. Talk about it with your doctor. Consider an increase or change in your medication. The perceived need for stimulation usually diminishes with proper ADD treatment.

Since we live in a Starbuck's sort of world, here are some practical guidelines.

- Small amounts of caffeine are not evil. One to three cups of coffee in a day won't undo the rest of your treatments. For the record, a cup of coffee is six ounces. The *grandissimo* size at your local coffeehouse might contain four to six times that.
- One or two cans of cola a day may be alright, too, but make it diet cola. Do we really need the empty calories?
- Mix decaffeinated 50-50 with regular coffee to immediately cut down the amount of caffeine in your favorite latte concoction. The Starbuck's guy absolutely can do this for you.
- Don't have caffeine within eight hours of bedtime. It may feel like it helps you get to sleep, but it doesn't, and you won't sleep as well.
- Learn to like other hot drinks. Green tea won't cure ADD, but it's so healthy for you that some nutritionists let you call it a serving of vegetable.

OK, let's get on to some bigger issues.

Cigarettes and other forms of nicotine are poor treatments for ADD. If you smoke, it's time to stop.

Nicotine actually treats ADD symptoms. If you put "stop-smoking patches" on hyperactive little boys, they will settle down and do their schoolwork. Nicotine is also a moderate anxiety reducer. There are two main reasons why we need *not* to smoke or use other forms of nicotine: first, there are better treatments available for ADD, and second, those treatments won't kill you early.

Many people with ADD have given up smoking only to restart later. If your ADD is properly medicated, there may be less inclination for you to restart[27]. One factor that feeds the desire to smoke can be eliminated—untreated ADD symptoms.

---

[27] A handful of patients have mentioned that amphetamine-based treatments (Vyvanse, Adderall XR and Adderall) increased their desire to smoke. There are no

There are many helpful stop-smoking programs available. See your doctor if you feel unable to accomplish this on your own.

There is very little good that alcohol can do in our lives and plenty of harm.

It is my belief that for the first two years after you are diagnosed with ADD, you should not have any alcohol. None. Zip. Nada. There is an important reason for saying this, so let me explain.

When you read that last paragraph, you had a response to it. I wrote that so that you could check your response, not because it is my real opinion. Maybe you're a tee-totaler and you thought, "OK, that was already my plan; no change needed here." Maybe you enjoy certain wines or mixed drinks occasionally and felt disappointment as if I had just told you to cut out desserts.

But the people this section is really meant for had a stronger reaction—either disbelief, anger or simply "No way." If that last paragraph upset you, consider that you may have a problem with alcohol. The problem is too big for this chapter to handle. The sum total benefit of all the advice in this book is not powerful enough to overcome the damage that alcohol can do to our minds. If you are an alcoholic or a problem drinker, your ADD will not get better until you deal with both problems together.

To those who truly drink little or none, just remember, "There, but for the grace of God, go I." Alcohol is potentially dangerous and absolutely not helpful in any way. You should understand the dangers clearly and respect this substance. Again, some practical guidelines:

---

good studies of this issue, so I asked Dr. Tim Wilens of Harvard University about this. Dr. Wilens is one of the most—if not *the* most—prominent expert in the world on ADD and substance use disorders. He knew of no studies reporting this association. Drop me an email if you know first hand about someone with this issue: survey@omason.com.

"Moderation" in a medical sense includes any alcohol consumption up to 2 drinks in 24 hours and 12 drinks per week. Some experts recommend those limits be halved for adults weighing less than 150 pounds. One drink equals a 12 oz. can of beer, a 6 oz. glass of wine or one shot of distilled spirits.

Since our brains are under a little extra stress, we will want more conservative guidelines than people with normal brains. I believe that 6 drinks per week is the upper limit of acceptable for people with ADD. (That is my own personal rule. I can't quote you a study for that one.) Remember that the mind making a decision after one drink is no longer the same mind that agreed to that first one; it is already less rational, more impulsive and measurably worse at driving home.

If you find yourself exceeding these limits even now that you know what is good for you, you may have a dependency on alcohol and should consider seeking help.

Keep in mind that the limits above are not the same as what is legal for driving. Some people will not be able to drive legally even though the medical profession says you were being moderate. We have enough accidents when we are sober. Research shows that alcohol impairs ADD drivers more than non-ADD drivers and that we are less able to determine our level of impairment than they are. There is simply not a *safe* limit that drivers with ADD can observe. Zero is our number if we will be driving.

In fact, zero is always the very best amount of alcohol as far as our brains are concerned. I have never yet lost track of how much I had, when I had zero to drink.

Marijuana and other street drugs are bad for all brains and desperately bad for ADD brains.

You already know where this is going, don't you? There is not much true "recreational" drug use. People with ADD who try marijuana, cocaine, street Ritalin or Adderall may do so in a recreational setting like a party, but if we continue using these, we do it because of the improvement we feel and the deep need we all have for symptom relief[28].

Marijuana is used regularly by many people with ADD, because it *seems* to help with ADD symptoms. This is largely a myth. Marijuana users feel as if they become more focused and efficient, but observers find that the users actually become less focused and less efficient. *Feeling* that you are more focused will not change your life. *Being* more focused will.

Illegally obtained Ritalin or Adderall are just not smart. It's much better to buy them legally and usually much cheaper, too. I've met many people who were amazingly good at "street pharmacology", but none of them knew enough that I would put my life in their hands to advise me on the issues of drug effects and safety. See a real specialist who can help you get on the right doses of the right medicines. Your brain is worth it. Imagine its potential when you treat it well.

[28] Nicotine is a legal drug that is used for similar reasons.

Chapter 7 THE CARE TEAM

We do best when we engage a team of ADD experts to help us reach our potential.

We've been talking about using medications, and it's obvious that we're going to need someone to prescribe them. It's time to think about the professionals that will help guide us through the steps of growth and change that we are anticipating with ADD treatment. There are a few experts that everyone will need along the way—doctors, therapists, coaches, organizational consultants and financial planners are the ones that we should consider early in our treatment.

We will need to find physicians who understand ADD and who are good at treating it.

It's not always easy to find a physician who treats adult ADD. Most of the physicians now in practice were not trained to treat adults with ADD, because adult ADD was not even understood during their years of schooling. There is not a Yellow Pages physician specialty category that lists all physicians with expertise in adult ADD. Let's take a brief look at the types of physicians that are most likely to treat adults with ADD.

- Psychiatrists. Adult and general psychiatrists are rapidly developing expertise and experience in treating adult ADD. In general, adult psychiatrists are the first specialty we should look to for help. That said, some have little or no experience with ADD, so you should ask. Some child and adolescent psychiatrists that treat children with ADD will also treat adults. Many won't, but it doesn't hurt to ask.

- Behavioral pediatricians. They are the sub-specialists who (along with child and adolescent psychiatrists) have specialized in the treatment of children with ADD. This recommendation might seem strange at first, but many have expanded their practices to include adults as well as children. Most behavioral pediatricians who have been in practice for more than a few years have had patients "age out" of their practices with no one available to assume their care. By default they have expanded into adult ADD care. Behavioral pediatrics is a subspecialty of pediatrics and many larger towns and cities have one or more available.
- Neurologists. Some neurologists do see ADD patients, but many do not. It is always worthwhile to inquire.
- Primary care physicians. This broad group includes family doctors, general internists, pediatricians (usually up until the age of 18) and many obstetrician-gynecologists. Experience with ADD diagnosis and treatment varies widely in this group. Those primary care doctors who do not diagnose ADD should be familiar with local resources and be able to help you find a specialist who does.
- Psychologists. Much of the work of diagnosing and counseling ADD patients for many years has been done by psychologists. Psychologists cannot prescribe medications, but many can diagnose ADD and refer to physicians who will prescribe them.

Once you find the physicians available to you, it is best to collect any references you can to help you decide whom to consult.

- Area CHADD groups (Children and Adults with ADD) can be helpful sources of information regarding all types of professionals who treat ADD. CHADD also maintains a database of professionals who treat ADD at their website: www.chadd.org.

- Another organization, the Attention Deficit Disorder Association (ADDA), maintains a similar referral database at their website: www.add.org.
- If you are seeing a non-physician therapist for ADD, it is often possible to obtain references from that therapist.
- Your family doctor will usually know physicians whom he or she personally recommends.
- School psychologists and social workers are often quite aware of local ADD care resources.
- Certainly, any friends who have ADD can relate their own experience with doctors whom they have consulted.

If your research leads you to more than one recommended physician, there is one other tip that might help you choose. Call the physicians' offices and identify yourself as a prospective patient. Ask very briefly: "Does your office diagnose and treat a *lot* of people with ADD?" If you don't get a clear 'yes' to that question, ask where they refer ADD patients.

It has been my experience that the physicians who treat a lot of ADD tend to have better outcomes and more satisfied patients. And certainly, we want to work with someone who has some professional enthusiasm about our case.

If you have been seeing a physician for some time, there is one more suggestion that may help you evaluate the quality of care you are receiving. Physicians who know ADD ask you for quite a bit of information to know how you are doing and whether further medication changes are warranted. If a doctor asks you how you are doing but doesn't probe further, you may not be getting the evaluation you deserve. The probing can be by means of a combination of direct interview, questionnaires, paper forms or even in the form of nurse interviews. Just know that it takes a lot of information to treat ADD. The doctor will ask *you* for the pertinent information; you won't have to evaluate yourself and guide the doctor.

Each case of ADD is unique. We need physicians who know our own specific case as well as they know the medical state-of-the-art.

ADD doesn't often exist in isolation. Four out of five people with ADD have some other mental health issues along with the ADD. We are much more likely than those without ADD to also suffer from depression, anxiety, dyslexia, bipolar disorder, learning disability, substance abuse disorder, addictions, sleep apnea, restless leg syndrome and a handful of other medical conditions. Identifying ADD is just a first step. Determining what accompanies it is equally important.

ADD patients don't exist in a vacuum either. We have family relationships, friendships and vocations that must continue as we are treated. Many decisions about treatment are modified by factors that research doesn't address. When does the swing-shift worker take his medications? What are the risks of medications during pregnancy? More importantly, what are the risks of stopping it during pregnancy[29]?

Finding the right physician to guide medication management is important in getting all the benefit available from medications. Since ADD medications are the foundation of therapy for most patients, the search is worth the time and effort it costs.

Skilled psychotherapists can make a huge impact in the improvements that you see with medication.

I hope I have been clear about the benefits of medications. They do not automatically force improvements, but rather they enable us to begin making changes. Therapists are experts who encourage us to make those changes and help us evaluate the roadblocks that hinder us. Typical topics with a therapist might include:

---

[29] For the record, there are no ***known*** risks to a fetus from any of the primary ADD medications. But neither are there any studies that demonstrate their safety. There are, however, known risks when women with ADD don't take their medications. Driving is less attentive and self-care is harder to maintain. This is a very tough quandary for those who conceive.

- Marital or relationship issues
- Feelings of sadness or depression
- Stress, anxiety or overwhelmed feelings
- A chronic sense of inadequacy or shame
- Parenting issues such as controlling our tempers with our children or disciplining reliably
- Past failures
- Healing when there is a history of abuse, loss, chronic stress, etc.

The following is an example of just one of the many problems that a therapist could help us work through, usually much better than we could on our own.

In most marriages where one partner has ADD, the spouse has accommodated for that partner's deficiencies. When the ADD partner begins medications, new problems may appear that neither anticipated. For example, a husband with ADD is often considered unreliable by his unaffected wife. (This is just a hypothetical case of a guy who isn't necessarily me.) Mr. ADD may have put considerable effort into trying not to forget appointments, with some success. Maybe when the two got married, he was late for every appointment. With effort and day planners and her reminders and his truly wanting to improve, he started arriving on time here and there. But being late for 8 out of 10 meetings does not yet meet most people's standards of reliability. It is still very frustrating to live with someone who can only perform at that level. So his wife developed an elaborate system of checks and balances to protect herself from the impact of his tardiness in her life. Consequently, she has grown accustomed to keeping two people on time.

So then, the husband with ADD is diagnosed and begins medication. Soon, things are improving right and left. He has much better time sense, and less forgetfulness. Without even trying, he is now late only 1 out of 10 times. In fact, he now apologizes when he is late, instead of acting as if late were normal and 'on-time' extraordinary. He is excited at his own improvement. Now he is spending less mental effort on timeliness and achieving far better results than ever before. Things go along pretty well for several days without any tardiness. Then once, just once, the husband shows up late for an appointment, and his wife hits the ceiling. He's hurt, because he's doing so much better, and she hasn't even mentioned it, but now that he goofed just once, she's more upset than she used to be when he was late all the time. What's going on here?

The story is quite different from his wife's perspective. Yes, she has prayed for years that he would become more responsible. When they were first married, she was burned too many times by his tardiness to leave it in his hands. She developed a finely-crafted system of prods and reminders to try to keep him organized and punctual. And yes, he is now much closer in this regard. His new 90% on-time performance may be a statistically impressive improvement compared to the original never-on-time rate, but the new 10% late rate is still enough to cause problems in social and business circumstances. He is causing her less embarrassment, but now it is no longer predictable. Before, she could reliably control his schedule. Now, he often does well without her efforts, so she dismantles her system and starts to treat him like a partner, not a child. He appears almost normal for a couple days and then—Wham!—out of the blue, he's not ready on time, and she is more ticked off than in the old days. She gave up the "control" of her old system, but the new system is still not as reliable as she needs it to be. He's getting treatment, so she is expecting "normal", but all she gets is "close to reliable". And "close to reliable" is not really reliable. It's going to take a whole new set of skills to live with her new, mostly improved husband.

These are common scenes in ADD households. Even the improvements that everyone hoped for can throw two partners out of kilter, especially when they have delicately balanced their life together around one person's disability. A psychotherapist who understands ADD would not only understand what is happening to this couple, but might have predicted it. Having someone with this level of understanding to guide us through the lumps and bumps of *improvement* can be invaluable.

The story above is about the more benign kinds of problems that need to be addressed in ADD therapy. Some much deeper problems require extensive therapy. Some marriages have been damaged by infidelity or failure to develop intimacy. Some families are broken and need extensive rebuilding. A high percentage of adults have developed unhealthy habits with alcohol or marijuana use that require specific intense therapy. In any case, there's a certain amount of catching up, improving or maturing that medications enable. Professional guidance and encouragement can truly ease the way, encourage hope and quicken the rate of improvement. And since there are probably many areas that could use improvement, a therapist may also be helpful in choosing and ordering the priorities for the specific areas in which to work.

Coaching helps build positive changes for many with ADD.

Coaching refers to a very specific kind of help that many with and without ADD have found very helpful. It is currently quite popular in business circles. Some executive coaches who are familiar with ADD have adapted coaching techniques specifically to help those with ADD, often with great success. ADD coaches are not therapists. Therapists help heal the past and figure out what isn't working; coaches encourage us and help plan the future. That distinction is entirely too simplistic, but it may help clarify the difference.

It is common to hire an ADD coach to help with a specific goal or problem area. Job issues are frequently targeted. Repetitive home or relationship problems are common subjects as well. Following is a list of just a few of the many issues that a coach might help us work through:

- Job searches or career changes
- Important projects at work, especially if they involve lengthy timelines or multiple stages

- Implementation or use of planners or organizational systems
- Chronic tardiness
- Home disorganization

An ADD coach first meets with you to help determine your goals. Together, coach and client then break the goals down into the individual steps that must be taken in order to reach them. The steps are fitted to a timetable. The coach checks in regularly—often over the phone or by e-mail—with the client to measure progress and fine-tune the timetable. Most people without ADD can accomplish these steps quite naturally in their own brains, but we with ADD often have to find ways to do the same thing outside our brains, placing task-orienting tools in our environments.

Finding an ADD coach will take a little digging. The Attention Deficit Disorder Association, www.add.org, and CHADD (Children and Adults with ADD), www.chadd.org, maintain professional directories on their respective websites. The ADD Coach Academy is an organization that has trained and certified numerous ADD coaches. Some have specialized interests such as coaching college students, executives, people with ADD and learning disabilities, and so forth. Their website is www.addca.com. The International Coaching Federation and the Coaches Training Institute train and certify many different types of coaches and can refer to ADD coaches through their websites, www.coachfederation.org and www.thecoaches.com, respectively.

There is no licensure or accreditation required to advertise as an ADD coach, so it is good to ask a prospective coach about his or her training and background. You will probably want to ask for references as well. Some ADD coaches without formal training or accreditation may be very gifted at encouraging, motivating and helping us plan; references and an interview may be the best way to evaluate someone without formal accreditation.

Professional organizers can be life transforming to the genetically disorganized.

Some people with ADD have managed to stay organized by sheer force of will and heroic effort. Most of us have always had messy desks and homes despite multiple efforts to reform. Some of us have organized spouses or co-workers who keep themselves and us organized and let us *think* that we're more organized than we are. Others of us have no organized helpers, and we will need to hire some help to get organized.

It sounds great to think that in a half day we could get our home office organized and truly usable. The same can happen for kitchens, closets and the garage. A professional organizer can transform a home or office in less time than it takes to read a book on organizing. Costs can run from $35 to $50 per hour or more, but the instant gratification may be well worth it.

There are two groups called NAPO, the National Association of Professional Organizers and POA, Professional Organizers of America, that are comprised of people who know how to declutter and organize. Referrals can be obtained on their websites: www.napo.org and www.nyorganizers.com.

You can also check www.organizeyourworld.com and www.organizerswebring.com for referral lists, or Google "professional organizer -software"[30] for more good ideas.

The "FlyLady" has a unique approach. Her free service aimed at homemakers will actually e-mail you reminders during the day to help you keep ahead of the clutter. Register at www.flylady.net.

We may need financial planners to help us get our personal economies on track.

Take a quick financial self-evaluation:

---

[30] Type the phrase without the quotes just as it appears on the page above. Adding the terms"-software" eliminates many references to organizing software which will clutter the search results.

- Do you have a will? Is it up to date?
- Are your credit cards all paid off?
- Do you have 3 months living expenses stashed away in an emergency fund?
- Do you have a budget and do you follow it?
- Do you have a method of adding to your savings at least every month?
- Do you know what you need for retirement and are you on track for acquiring it?

Very few people with or without ADD can say yes to all these questions. If you can, you don't need to read the rest of this chapter. In reality, though, most of us are not doing as well financially as we should be. Adults with ADD average $10,000 per year less income than those without. Credit problems, bankruptcies, late or unpaid taxes, overdrawn checking accounts and maxed out credit cards are common problems for very many of my patients. Even if financial bombs aren't bursting, we still live better if our account books are in order. Virtually everyone without ADD needs professional fiscal advice to make that happen, and so do we. Almost all of us will need to hire experts to help us get our finances organized.

One needs to be careful seeking professional financial help. It is best to get financial advice from someone who has *your* best interest at heart. Lots of people want to help you reorganize your money in a way that will definitely benefit them, but not necessarily you. Many who offer financial planning services also sell investments, insurance, loans, etc. A "planner" who is itching to get to the part where he earns a commission from you is not who you need right now. There are currently hundreds of shady financial planning services, get-out-of-debt services and the like which are really fronts for high-rate loan companies. So don't Google "financial planner". You might as well Google, "take the last of my money, I misuse it all anyway".

The type of help you need varies by your current financial situation. If you are financially drowning, the steps to take are basic: create a budget, pay off debt and begin saving. You probably can't afford to hire someone to help you at this stage, but there are good alternatives.

Credit counselors help people get out of debt. Many of these work at subsidized rates through banks, local government, or neighborhood organizations. The FTC has solid advice on choosing a credit counselor; search "choosing credit counselor" at www.ftc.gov.

There is an excellent seminar series called Financial Peace University (www.daveramsey.com) that is often taught in churches, but is also available in video or book form. It is available both with and without a faith-based emphasis. The financial advice Dave Ramsey gives is right on.

Mary Hunt's "Cheapskate" series is another great place to begin. It includes seminars, a monthly newsletter (www.cheapskatemonthly.com) and numerous books (e.g. *The Complete Cheapskate: How to Get Out of Debt, Stay Out, and Break Free from Money Worries Forever*). Mary was a credit card addict who amassed over $100,000 in credit card debt before she figured out that financial chaos is definitely *not* fun. Even if you're not financially drowning, these series have a lot of generally useful advice about living within one's means.

Those who aren't in a financial straightjacket, but still don't feel in control of their finances may benefit from a financial planner. "Fee-only" financial planners charge a fee for their services and do not try to sell you anything. You can contact the National Financial Planning Support Center at www.fpanet.org to find a Certified Financial Planner to help guide you. (Not all Certified Financial Planners are fee-only, though.)

Accountants often provide financial planning services or can refer you to a financial planner as well. Suze Orman and Andrew Tobias are two personal finance authors that dispense sound financial advice, and they both write well enough that ADD don't-like-to-read-ers might even enjoy reading them.

It's time to assemble the team.

Who has the time and money to hire all these experts? Certainly not most folks. But don't simply dismiss this advice because it seems overwhelming. Not all of the team needs to be assembled this week. We all definitely need a doctor, and most people with ADD truly benefit from a good therapist. The other consultants need to be considered in light of the misery they can save you. What would make life better—a new iPod or an organized home? And how much are we spending to support our financial chaos? Add up the last year's interest on credit cards, late fees on bills, and higher interest rates on auto and mortgage loans due to an imperfect credit rating. Professional financial help may be cheaper than what some of us are already paying for the privilege of mismanaging our own money!

It is always tempting to buy the self-help books for $15 or $20 instead of hiring the professionals at $50 per hour. For some people that may be a very wise choice, especially if you have a partner who will work with you to implement change. But you need to know yourself in this regard. If you might buy a book and even read it all, but are not likely to do the work of getting organized, it is better to hire the professional and actually get it done. Remember Yoda's advice to Luke Skywalker: "Do, or do not. There is no try."

Everyone I've met with ADD has already tried extremely hard to get organized, repair relationships and make things work. Most are exhausted from all the trying. It's amazing how many keep valiantly trying when trying has failed so many times. And all our well-meaning friends say to try harder, so we do. But if trying harder is not getting it done, I'm giving you permission to stop trying, raise the white flag and call in the professionals. Yoda would be proud. More importantly, so will you.

Chapter 8 GETTING ADD "SMART"

Getting better requires learning all we can about ADD.

Most of us with ADD can remember a time when we assumed we were "normal". Then at some point it began to dawn that we might have ADD. Someone who had seen it before took a look at us and said something like, "You know, you have a lot of similarities to ________", and of course the person they compared us to also had ADD. The remarkable thing is that there were so many years prior to the truth dawning. Those are the "years without awareness". Let's have no more of those.

ADD is hard for us to suspect in ourselves, because it impairs our ability to self-observe.

One of the brain's important functions is the ability to watch our own actions and effects in the world and continually adjust our efforts to make them more effective. It's almost as if the mind contains an impartial observer that is continually giving feedback on our progress. If a friend becomes bored in a conversation, the evaluator notices it, and we have the opportunity to stop talking and ask a few questions…unless we have ADD and don't notice stuff like that. How can we notice that we're bad at noticing, if our 'noticer' is broken?

So, untreated, it's hard to notice ADD traits in ourselves through the fog of inattention. With treatment, medication improves a brain function known as "self-monitoring", so we actually improve awareness of our own ADD symptoms. I've had a few patients who felt as if they were doing *worse* after starting medication, because they had become much more aware of their forgetfulness and disorganization, even though their symptoms were much better in the eyes of their families.

Basically, it's not good to have a disability and not realize it.

Pretend for a minute that you were born without two of the fingers on your left hand and no one had ever noticed it. As a child, you would not be as capable with that hand as other children, and it would cause problems at times. You would struggle making a cat's cradle string figure. Your Thanksgiving handprint turkeys would look wrong. Baseball gloves wouldn't fit so well.

Your parents would no doubt have tried logical steps to help. They might, for example, enroll you in a guitar class to improve your dexterity. Sadly, you would have entirely disappointed yourself and your guitar teacher. (Most chords are played with three fingers and the thumb.) Your failures would have further cemented in your mind that you are not very good with your hands.

Now let's say that one day your guitar teacher is watching you fumble and struggle, and it suddenly strikes her that your left hand is different than other students'. She speaks privately to your parents and recommends they take you to a hand specialist, and finally, the truth becomes clear. You lack two fingers.

The specialist's discovery is going to revolutionize a lot of things in your life. Your guitar lessons will need to change. Either they must be adapted to your disability or you will probably quit them. You might decide to take up an instrument that isn't dependent on fingering like the trombone or drums. (My roommate in 8th grade band camp was born without fingers on his right hand, and he was a gifted trombonist.) Finally, if the world were truly just, your parents might even apologize for their unrealistic expectations over the years. But whether you decide to continue adaptive guitar lessons or tackle a different instrument, you have at least identified the challenges involved when you make the decision.

This absurd example is instructive. It is ludicrous to expect certain functions of someone who is physically unequipped to perform them. Anyone who has ever made an unthinking comment to a disabled person has faced this. In college I once said to a blind friend, "You should see what some kids are doing in the library." He started laughing, but it took me a few embarrassing moments to figure out why. I had said something so ridiculous that he either had to laugh or hit me. Fortunately, he was gracious and laughed.

When our ADD was undetected, there was not much laughing. People said things like, "You just need to try harder," when we were giving everything we had. They said, "You just don't care enough to learn this," when, in fact, we did care and might have been better able to learn if the lesson had been taught differently. Inside, we started believing what others said. Somehow it was easier to believe we didn't care than to admit we were incapable.

When ADD is diagnosed, we finally have the chance to get comfortable with ourselves. We begin to see exactly where ADD has most impacted our lives. The past starts to make sense. We realize that certain habits and compulsions are adjustments to compensate for ADD symptoms. Other patterns which we thought were character flaws turn out to be common ADD symptoms. Once we understand the characteristics of ADD, our own life patterns become easier to understand.

The disability of ADD differs from most other medical conditions.

ADD is not the easiest disorder in the world to pinpoint. Skin conditions probably are, and sometimes I envy the dermatologists whose consults often take five minutes or less. In contrast, every diagnosis of ADD takes several hours of the combined efforts of a patient, his or her family and the medical team.

The main reason doctors have to look so thoroughly to diagnose it is that ADD—unlike most diseases—does not cause unique symptoms. Psoriasis has a unique look and is quite different from normal skin. People with epilepsy have seizures; the rest of us don't have them. People with strokes lose control of one half of their body; people who aren't having a stroke don't have that symptom. In contrast, we with ADD lose our keys; well, so does everyone else at times. We lose focus during boring lectures; so does everyone else, just not as often. No unique thing happens that conclusively signals the presence of ADD. The symptoms of ADD are disabling because of their *high frequency*, but they are not extra-ordinary or unique like chest pains or enlarging moles.

A second reason we have to work so hard to diagnose ADD is that it doesn't suddenly appear. It has always been present. Most diseases, in contrast, represent a change from a lifelong baseline. Strokes are an example of this; they affect people who have enjoyed many years of full body control, until suddenly everything changes. Pneumonia makes formerly healthy people feel miserable with cough, fever and chest pain that can develop over just a few hours. People with most medical disorders come to the doctor and describe a change from their baseline. We with ADD have no change to discuss. We were always more or less this way.

The final reason ADD is hard to diagnose is that it doesn't always look the same. Some of us have hyperactivity without attention problems, some of us have inattention without hyperactivity, and some of us have both. (Many teachers have told parents that their inattentive child does not have ADD, because he or she lacks the activity level of the "classic" cases.) Most of us haven't done well in school, but some have. Substance abuse, depression, auto accident and crisis pregnancy rates are much higher for people with ADD, but many of us have had none of these. There is no single factor that can certify that a person does or does not have ADD. The diagnosis of ADD requires listening to enough details that a pattern emerges.

Since ADD is tough for even professionals to identify, we will have to devote significant effort to seeing it in ourselves.

We can figure out how to live well with ADD when we understand how it manifests in us. When we run into roadblocks, we need to know whether it is an ADD problem or something else. ADD can look like "lazy" or "not trying" when nothing is farther from the truth, but then there are some times when we are frankly lazy and should be trying harder. Blaming our laziness on ADD is not going to help anything. All the Ritalin in the world won't get us up off the couch and start our chores for us.

How can we know whether a roadblock is due to brain disability or some other factor like laziness? There are two test questions we can ask.

- Have I tried my honest best to overcome this problem using the methods that work for most people?
- Is this a common problem for others with ADD?

Question 1 is answered with self-searching. Question 2 cannot be answered unless we are pretty familiar with ADD. Martin Luther King Jr.'s words are particularly apt:

"Education is the key to knowledge and knowledge is the key to power and freedom."

Each of us needs to learn how ADD is personally affecting him or her.

Conferences, seminars and self-help groups are often good sources of information. (A variety of conference information can be found at www.chadd.org and www.add.com). People in larger towns and cities probably have local CHADD chapters that meet regularly. ADDA and CHADD both have excellent national conferences with lectures and workshops on a wide variety of ADD topics. Both organizations have regional conferences, too. And nothing beats the relatively low cost and rich returns of reading good books. Education for most of us will include some reading.

Now I sincerely hope that the reading you have done this far in this book has helped you understand ADD more. But please know that this book gives only an introduction to many complex issues. Other smart people and skilled writers have gone into depth on these issues in ways that you may find far more useful than this. I am going to recommend a number of these that my patients and I have found routinely helpful. This list is by no means complete, but it is at least a starting point for learning.

Not all of us are readers. Most people with ADD struggle to read and don't enjoy it. But reading to learn can be done in ADD-friendly ways. It is not necessary to read any of these books cover-to-cover. They have value even if you pick them up and just read random chapters all out of order. And many are available in books-on-tape form, if that is a better format for you.

Books to help learn about ADD in general

Driven to Distraction is the classic in this category. Drs. Ned Hallowell and John Ratey published this in 1995, and it has been helping to change lives ever since. *You Mean I'm Not Lazy, Stupid or Crazy?!* by Kate Kelly and Peggy Ramundo (1993) is also written to explain ADD, and it is written in a style that many with ADD find easier to read. Both books are a bit out of date, especially in their discussion of medication treatments, but they are great at helping us learn what is and isn't ADD. Dr. Hallowell published *Delivered From Distraction* in 2005, and it is more up-to-date, but I'm not sure that it is better for our current purpose.

ADD Success Stories by Thom Hartmann has been helpful to many trying to understand this disorder. It has a quite easy-to-read format with many short stories from the lives of others with ADD. It's pleasant to read, because the emphasis is on successfully dealing with this disorder.

Joel Young MD wrote *ADHD Grown Up: A Guide to Adolescent and Adult ADHD* which is a good, broad review of adult ADD, and I would plug his book even if he weren't a friend of mine. *Attention Deficit Disorder: the Unfocused Mind in Children and Adults,* by Dr. Tom Brown, is a very in-depth book for people who like science and want to understand how the brain works. *Scattered Minds: Hope and Help for Adults with Attention Deficit Hyperactivity Disorder* by Dr. Len Adler and Mari Florence is another authoritative work that should be on your short list.

Books about ADD relationships

What Does Everybody Else Know That I Don't? Social Skills Help for Adults with AD/HD, by Michelle Novotni, is a great place to start. We'll discuss it in more detail in the next chapter. Kathleen Nadeau goes into depth about marriage, family and relationship issues in *Adventures in Fast Forward.* Another highly regarded choice is *Is It You, Me, or Adult A.D.D.? Stopping the Roller Coaster When Someone You Love Has Attention Deficit Disorder* by Gina Pera.

Books that are helpful for other common ADD problems

Not every problem we need help with is unique to ADD. Many people struggle to get organized. Clutter affects many homes, not just ours. Proper diet, good sleep habits and regular exercise are struggles for almost everyone. The following books are not ADD-specific, but often helpful nonetheless.

Since we mentioned clutter, let's go there. There is an outstanding series of books on that topic by Don Aslett, beginning with *Clutter's Last Stand.* They are written with brief chapters, humorous examples and a motivating style. His emphasis is on living happier with less clutter and less dirt.

There are other books that emphasize how to organize what we have left after de-cluttering. Donna Smallin is a friend and a professional organizer who wrote *The One-Minute Organizer Plain & Simple* which focuses on this topic. *ADD-Friendly Ways to Organize Your Life,* by Judith Kohlberg and Kathleen Nadeau, tackles clutter, organization *and* time-management issues. These are just three examples of many helpful titles on the subject.

Finances are a struggle for most of us. *Financial Peace: Restoring Financial Hope to You and Your Family* by Dave Ramsey is a good place to start. It's not about getting rich by playing the markets. There's no discussion of which mutual funds to buy or how to select stocks. It's all about living within our means, no matter what those means are. This approach is vitally important for all of us with ADD. There are a few of us who have made a lot of money, but far fewer who have kept much of it. The goal in money matters is not to become so filthy rich that it's impossible to overspend our income, because no such fairyland exists. (The US government can't even get by on the measly 2 or 3 trillion dollars we send them each year.) Rather, we need to learn how to responsibly enjoy whatever is available.

There are thousands of books about self-actualization, many of them with a professional or business slant. *The Seven Habits of Highly Effective People*, by Stephen Covey, is a classic that is well-worth reading. These habits are important to everyone, not just those of us with ADD.

My next suggestion takes a wild tangent, but humor me for a minute. There is a book on gardening that struck me as quite educational. It's *The Impatient Gardener* by Jerry Baker. He emphasizes doing simple things that yield big results. He explains how small tasks timed properly can save many hours down the road. It's a helpful perspective for chronic procrastinators, and it got me thinking far beyond just gardening, even before my ADD was diagnosed. That said, don't buy it if you don't own a garden.

Finally there are all the other self-help books in the world. Any of them may be useful to some of us, since ADD affects all of life. Heck, there's probably an ADD-friendly, vegetarian cookbook that could help us improve a bit. But for our purposes now, we want to figure out ADD and the ignorance that keeps us from seeing our problems accurately and solving them successfully. It's probably better to stick to a few basic books that we might actually finish than it is to list 30 great ones to ignore and feel guilty about.

We need to learn about ADD so that we can see ourselves more accurately.

Many of us with ADD are afraid to look closely inside. I believe that is because we fear that we are going to find distressing things when we do. What if we discover laziness, indifference, poor motivation, weak character or incompetence just as others might have suggested to us over the years? Unfortunately, we will find some of that. But we will also find that some—maybe much—of what we thought was miserably flawed character is simply a disability called ADD. The disability need not be feared; it can be treated with medication, counseling and personal growth. And once it is treated, it is much easier to attack the real flaws inside us—the laziness, greed or selfishness that are not themselves a result of ADD.

Most importantly, we can learn to see the difference. Wouldn't it be freeing to learn that? I believe it would.

Martin Luther King Jr. thought so, too.

"Education is the key to knowledge and knowledge is the key to power and freedom."

Chapter 9 LEARNING HOW TO TREAT OTHERS

Maybe we should go to finishing school.

Remember finishing school? When I was younger, I couldn't believe my ears when I first heard about the poor, unfortunate girls who had to endure that. Tea parties, table settings, curtsying and practicing polite drivel—and their parents *paid* to make them do that! What warped sensibilities!

Forty years later and from a different perspective, I am beginning to see just a little wisdom in that approach. There are many folks who could benefit from schooling in social skills, including many of us with ADD.

People with ADD may lack social graces.

Was that gently put? I wanted to come right out and say that a lot of us with ADD are raging social misfits, but I'm trying extra hard right now to be sociable and polite about this topic. Blurting the truth without sugar-coating it is one of many common foibles we are known for. A small child may loudly announce things like, "That man only has one arm," and we all blush for the man and the child. The little child was only stating the obvious truth, but everyone older than seven has learned that "we just don't do that". There are some straightforward social conventions that we were all taught as children—including 'speak privately when talking about others' and 'don't point at people'. These are but two examples of a distinct list of social conventions that we were explicitly taught.

But there is an even longer list of social conventions that we were probably *not* explicitly taught. How far do you stand from a new acquaintance when conversing? How do you convey regrets to a coworker on the death of a family member? A neighbor? An acquaintance at a club? When is a hug needed, when is it a good idea, when is it inappropriate? Which polo shirts are acceptable "business casual" and which are not? When is touching someone's arm that you are speaking to good, and when is it not? Life is full of nuanced circumstances that require that we a) know "the rules" and b) know how to adjust and apply them on the fly. Figuring these out requires a degree of attention that many with ADD are not able to give.

I have known professionals with ADD who wore basketball shorts and old running shoes to work, or who didn't shave or brush their hair before seeing clients. It is common for people with ADD to test the limits of whatever explicit dress code is in effect. There are many reasons given for this. Some despise conformity and choose to flaunt the rules. Some are still struggling with morning routine and don't have time to do better. Some have misplaced what they had hoped to wear and had to make a hurried selection at the last minute. But the curious fact is that many with ADD, for whatever reason, dress at the limit (and sometimes beyond it) of what is considered acceptable. This is just one example of a social skill that is commonly weak in those with ADD. Some individuals with ADD have also been known to:

- bump others without saying "excuse me" (or even realizing we did it)
- interrupt conversations in the haste to say something before it is forgotten
- talk too long about a subject and miss all the cues to stop
- focus on ourselves or our own needs instead of others
- blurt out something only to then realize after saying it that it was hurtful
- forget the tastes or needs of our friends, even after several reminders
- go on and on relentlessly about other people's foibles, as if the rest of us even cared about their

> thoughtless rantings, ignoring the growing discomfort they are creating and not only that, but…

Oops, I better stop already. You get the picture. Finesse is not our strong point. Bluntness seems to be where we excel.

Social finesse is a valuable skill worth our focused efforts.

Social skills are effective in our world. Lacking them hinders success. It is better to learn them late than never. We need to get to work and upgrade our skill set. The cost of *not* doing this is high. Most people with ADD have a deep, heart-felt concern for others. But having these feelings doesn't help others until we can communicate them tangibly. Love isn't love until it is shared, and it is generally best shared in the language of the other person. We have to transfer it from inside of us, over to the people for whom it is felt.

Let's be real, though. Much social finesse begins with noticing the people around us and their needs. Our noticing mechanisms are congenitally deficient. How are we going to improve?

The answer is simple: we get better by working at getting better. There is no more effective option. Medication may help the noticing mechanism, but it won't say "excuse me" for us.

There is a simple program for improving social skills that we all can begin.

Step number 1: Find a real friend/accomplice who agrees to give you honest feedback. This person needs to be able to watch you in a variety of situations and be willing to schedule time with you to talk about how you are doing.

Step number 2: Buy *What Does Everybody Else Know That I Don't?: Social Skills Help for Adults With Attention Deficit/Hyperactivity Disorder* by Michele Novotni. It's a book about precisely this topic. She goes through a lot of unspoken social conventions that can smooth the way through the world.

Step number 3: Start the program. Read a new chapter in this book every month or two. Meet with your friend/accountability partner and set some realistic goals for improvement in the next month. Try to frame the goal in terms of something that you can measure. "My goal is to be really nice this month" is hard to quantify. "I will maintain eye contact when we talk" is much easier for this friend to assess. Agree when you'll meet again. Adjourn.

My brother and I tried this and—because we live 5 hours apart—we did it by email. For example, he was working on focusing on others and not on himself in conversation. His job at a retail store gave him many brief encounters with customers. We set a goal together that he would count how many times he could respond to a customer without using the words "I" or "me" in the sentence. He would track himself and email his score to me every evening. By the end of one month, his scores had dramatically improved. More importantly, his focus on a customer-service skill led to improvements in his friendships and family relationships as well.

Select the social skills that need the most attention.

What are the kinds of social skills we should work on? Obviously, each of us needs to research this individually, because each of us have somehow picked up some of the skills and missed others. I've included my own Top Ten list of some social skills that most of us should consider polishing. It is by no means exhaustive, nor applicable to everyone.

Top Ten List Of Social Skills We Should Consider Improving

10) **Dressing for success.** There is a tendency among people with ADD to dress at or even below the lower edge of acceptability. Fortunately, we live in a moderately tolerant society. But there is some self-centeredness in the assumption that others must look past outward appearances to see the real person inside us.

9) **Good grooming.** This is kin to number 10. It might be disappointing that the world is full of people so shallow that they actually care whether our hair is neat and our clothes ironed. But rumpled clothes and "bed head" scream out "Don't take me seriously, because I certainly don't."

8) **Brevity in conversation.** Like this.

7) **Remembering names.** Very few people with ADD have a gift here. On average, we are pretty bad at it for an obvious reason: forgetfulness is part of ADD. But remembering people's names and using them is a sure way to help people feel good about an interaction with us. It's a skill worth cultivating. When I was in college, I met Dr. John Alexander, who was the president of the Inter-Varsity Christian Fellowship. Six months later, I ran into him at a conference of over 20,000 people and he immediately remembered my name. Not only was I amazed, but I felt very affirmed. Later, I asked him how he was able to do this. He explained that he had always had a gift for this, but that he made a habit of repeating the name of a newly-introduced acquaintance three times to himself before the conversation ended. It's a good habit to practice, even if we don't have his gift.

6) **Driving considerately.** Yes, some people drive like idiots. Yes, some people are infuriatingly rude and thoughtless. But that will always be. We don't have to respond in kind. It accomplishes nothing good and is potentially dangerous. Furthermore, let's be real. How often is the true reason for our reckless hurry our own poor planning? It is actually not the fault of the idiot driver in front of us when we feel the need to speed. The roads are plenty unsafe without us adding to the danger. Drivers with ADD have several times the accident rate of those without. Let's consciously do all we can to reduce that. Let's leave on time when we can, call ahead when we can't, and drive safely in all cases.

5) **Showing interest in others.** Many of us have developed habits of joke-telling, clowning around or story-telling as substitutes for true friendliness. These are often poor substitutes that may leave people feeling entertained when the encounter is over, but not connected or valued. Michele Novotni suggests a habit in her book that I think is brilliant in its simplicity. In conversations with people, she recommends asking 3 different questions about the other person and pursuing each with two more follow-up questions. The technique is called "Three questions, three deep", and it is a good way to practice showing interest in others. I suggest this, not because there is much good in faking interest in others, rather because I believe that most of us with ADD have far more regard for others than we have actually yet shown them.

4) **Apologizing.** Everyone, even Ms. Manners, bumps into others in a crowd or makes inadvertent blunders. The proper response is a simple "I'm sorry." That's all that needs to be said 99% of the time. It really helped me to learn that an apology is not an admission of mean intent. It's a simple statement of empathy. 'I feel bad that __________ happened to you. I'm sorry.' It doesn't mean we're necessarily in the wrong; it means we care. Practice saying it when no one is listening and offer it freely whenever it might apply.

3) **Giving space.** Sometimes the best way to befriend or love others is to back off. When we apologize for hurting or disappointing someone, they might still need time to let it pass. We might want the bad feelings to quickly blow over and everything to return to normal, but that person needs some time to recover, and we have to quiet our impulsive, pressing desire to fix things *now.* Or maybe my work schedule is so light that I can kibbutz, but my coworkers can't right now, so they rebuff my overtures. It's not that they dislike me, they are just too busy. It's not personal, so there is no need to feel badly, but it's important to recognize and honor their needs.

2) **Reading body language.** People are constantly giving us valuable, non-verbal feedback, but we have to notice it. I have a history of being very, very bad at this. When I have tried to get a point across in conversation, I might see someone look away or look disinterested and would assume that I was not making my point well, because the point was certainly interesting to me, not boring. So I would press the point harder, try it again from a different angle, keep going with it. The predictable result was that people would write me off as an opinionated fool, rescue me by creating a diversion, or escape with a mumbled "That's interesting" which happens to be the socially correct lie most people have learned to use when they are anything but interested. Actually, I was missing very clear body language clues that I was wasting my words. People were using Standard American Body Language to tell me to "shut up". And I wasn't paying attention. If our words are boring others, they will start to avoid eye contact and shuffle distractedly. If we are making others defensive or angry they may cross their arms or legs. If those are the results we intended, well and good, but if not, they are telling us it's time to shift tactics and we would do well to "listen" to what their body language is saying. I never really noticed that people telegraph their needs until after I was on medication.

1) **Sensitivity to others' feelings.** This is the highest and hardest of all these skills. It's so important that I'm going to give it a more thorough look in the next section.

Increasing our sensitivity to the feelings and needs of the people close to us is worthy of our highest efforts.

Biology probably makes "emotional listening" even harder for men than for women, but it is a learned skill in either case, and all of us can learn to do it better. For example, if Jennifer just told her husband-with-ADD, Joe, that she had a lousy day at work, what should he say?

a) "Are you done, because I need to know if you'll have dinner ready in time for me to get to softball?"
b) "That's almost as bad as my lousy day…"
c) "I told you they're all miserable slimeballs at your office. When are you going to quit?"
d) or "That sounds hard after all the effort you put into that project. I've got time if you'd care to tell me more about it."

(For any Neanderthals who answered "a", I'll just point out that "d" is actually correct.) Offering to listen may be the last thing that comes to Joe's mind, but it may be the thing most likely to help Jane feel better, put the bad day behind her and be more attentive to the family for the evening.

On the other hand, sensitivity to feelings may be somewhat more innate for a woman, but effective responses to her husband's feelings may not be so innate. At another house on the same block, Chuck comes home and says, "Boy, what a lousy day at the office!" should Cathy, his wife-with-ADD say:

a) "It's not as bad as my day chasing after our ADD kids."
b) "Maybe you want to borrow my medicine tomorrow?"
c) "That sounds hard after all the effort you put into that project. I've got time if you'd care to tell me more about it."
d) or "You must be tired. Let me rub your shoulders. Maybe we can get the kids to bed early tonight, then we can build a fire and have a great evening."

Surprisingly, "c" is not the correct answer. Chuck, it turns out, is not the sort of person who feels better rehashing a bad day. Rather, he would feel better, perk right up and be far more attentive to the family if Cathy chose "d".

Sensitivity is a learned response.

Some people (like Joe in the first example) naturally want to fix things, so they respond with advice or evaluation when listening and empathy are needed. Jennifer needs to be heard. Joe needs to make intentional efforts to learn to read his wife's feelings and respond to them. Some people are good at understanding and empathy and may be quick to offer those. For people like Jane, they are helpful, but for others like Chuck, they are not what is needed. For Chuck, diversion and "being taken care of" are the language of love. Cathy needs to learn that doing what would feel good to her might not be the preferred response for someone else whose brain is wired differently.

So what is the ADD connection here? This sounds like generic "men are from Mars, women are from Venus" advice. And that is correct so far. Here is the main point for us with ADD. Being this sensitive, this understanding—for people with ADD—is literally heroic. Cathy, who is exhausted from chasing her hyperactive kids all day, has to reach far outside her own needs to do what is not natural but will lift Chuck's spirits. Joe, whose disorder includes "has trouble listening when spoken to directly", has to reach far outside his natural abilities to offer to listen to Jennifer when she most needs it.

Many of us with ADD have poor listening skills and are embarrassed to offer them, but the need for that skill is so high in most close relationships that it is usually best to just put our feeble skills on the table and start practicing and improving them. *The Five Love Languages,* by Gary Chapman, may be helpful reading for anyone who wants to explore this specific topic.

In conclusion:

We've touched on everything from simple rules of etiquette, like dressing appropriately and apologizing, all the way to subtle skills in intimate relationships. No one with ADD lacks all of these skills, and no one without it does them all perfectly. They are simply the tools that help us get along and succeed with people. They don't begin to happen until we begin to pay attention to others.

For myself, ADD treatment probably hasn't given me more care for other people; rather it has helped me to show my care more. That is the biggest blessing of the treatment for me—to show more love. Isn't that what makes you happy, too? It's not about straightening our desks—it's about each other.

Thank you for reading and considering this. Really. Thank you very much.

Part 3

Long-Term Strategies

Chapter 10 FITTING IN WITH THE PEOPLE IN OUR LIVES

To encourage our best potential we should surround ourselves with people who wish us to get better.

We turn now to longer-term strategies for living with ADD. There is no "to-do" list that you will have at the end of these chapters. Rather, I hope they give you a mindset that helps make living with ADD better.

Having ADD is a basic part of our personalities, and treatment is not going to change some fundamental aspects of who we are. Treated or not, there are some patterns in our relationships worth understanding.

Let me add a disclaimer to this and the following chapters. No single observation or bit of advice applies to *everyone* with ADD. I'm going to make some sweeping generalizations which are often—but not always—true. There is not much scientific data behind most of what I'll say from here on; it is rather the collected wisdom and observations of my patients and colleagues.

The right friends can help us get better.

Most people develop very few true friendships in a lifetime. We call people 'friends' who are actually close acquaintances. When we say "my friends", we usually mean "the group of people that I get together with more than once per year for a meal or a party." When I say "your real friends", I'm referring to the entire group of people who will shave their heads when you start chemotherapy, just so you won't have to be the only bald one. So let me say it again: most people have very few real friends in a lifetime. But just one such friend could make a world of difference in our life journeys. Real friends might just tell us what we really need to hear.

Since listening skills are something that most of us need to work on, let's consider how to enlist a friend's help. Someone who hopes to become a better listener needs to clearly invite his friend to offer that critique. (After all, if the friend doesn't have ADD, he will probably withhold most criticism, simply because that is our social custom.) Then the friend can remind him when he did not listen and encourage him when he did it well. That sort of monitoring and feedback could help improve not just listening skills, but eye contact, punctuality and project completion. A real friend will care enough to be honest and understand us enough to do it kindly. Over time, the cumulative effect of friends' love is life-changing.

We need to be around people who value our strong points. Most of us have not just glaring weaknesses but glaring strengths, too. Many with ADD have qualities like spontaneity, humor, novel insight and out-of-box thinking[31] which are great assets at a weekend party. Qualities that had to be bottled up at work all week can be perfect at social get-togethers. A spirit of adventure is common for people with ADD and might counterbalance the reticence of a friend who is more conservative. It is healing to find people who value us. People who value our strengths have a powerful positive effect on those of us whose self-esteem is chronically low. Friends who can forgive our faults provide powerful motivation to overcome them.

Finally, we need to be wary of friends who drag us down—keep us from growing and getting better. When we begin to avoid certain risky behaviors, those who undermine our intentions are not good to have around.

---

[31] Again, these are not ADD-specific characteristics, per se. Others with ADD don't possess these qualities, but will have different strengths.

Those who are recovering from drug or alcohol abuse face these issues squarely, but it is equally important for people who are trying to drive more safely, snowboard more responsibly, arrive on time or quit gambling. "Friends" who urge you to do what you have decided not to do are not your friends. We need to nurture relationships with the people who help us become better.

Conversely, getting better can help us be a better friend.

People who fail to listen well, blurt and interrupt, show up late for meetings, and who think about writing letters but neglect to send them have a bit of trouble cultivating friendships. Talking too much and failing to listen can drive people away. Learning to listen, though, makes friendship much more likely.

It's been said that "Friendship isn't a big thing—it's a million little things."[32] Treating ADD helps us do the routine things that maintain friendships day after day, so that we have friends already available when exciting, compelling times happen to come along.

The right coworkers can help us achieve professional success.

Some people with ADD have made their way successfully in their professions and work life, but many have struggled below their capabilities. It is not uncommon for me to meet a patient with ADD who has been fired or forced to quit more than a dozen jobs. There are many things that contribute to success in the working world; three are especially important for people with ADD to consider.

First, as far as possible, choose the right boss. There are some who cannot work for any boss but themselves, and further advice on "working together" cannot be helpful. These folks need to run their own business. Most with ADD, though, are

---

[32] Author unknown.

not in a position to start or run a business and will always have a boss. For the rest of us, then, there are qualities in a supervisor that will dramatically improve our performance.

Whenever there is an option, we should strive to work for bosses who a) value creativity over conformity, b) set firm goals and clear deadlines but offer flexibility in how they are accomplished, c) have high expectations and d) motivate with encouragement, not criticism. It is almost always disastrous when someone with ADD tries to work for someone humorless, who values conformity and criticizes freely. I know this; I've tried it, and it was very ugly. Please save your self-esteem, and keep looking.

Second, we need to know ourselves well enough to know whether we thrive in teams or alone. Some do better with the camaraderie and *esprit de corps* of a tightly-knit team. Others simply do best when they are flying solo and everything hinges on them alone. If you work better on your own, it is best to aim for positions in which you are not on a team and vice versa. To do otherwise means always swimming upstream.

Third, we need people in our work world who can finish what we start, wrap up the loose ends we leave, dot our i's and cross our t's. There are people who are great at details. They might have a "just business" attitude that can even be off-putting on a first meeting. Their desks are neat, their affairs are in order, and they hardly smile at our jokes. The last thing you'd think when you meet such a person is "Gee, I'd sure like to go out after work with him/her."

We don't have to hang out with such people, but we desperately need them to work with us. They are the yin to our yang. They are everything we are not and vice versa. We need them as coworkers on our teams, and if we supervise others, we need them working for us. We start the work; they make sure it gets done. We have the novel ideas, they execute them. Detail people increase our chances for success. Simply finding the right people to work with can improve our performance.

We need to understand the impact our life partners can have on our ADD and that our ADD has on them.

Dr. William Dodson is a psychiatrist in Denver who exclusively treats adults with ADD. He has observed that there are three typical partner choices that his patients make. Many marry someone with the skills of an executive secretary. This is the person whose life is neat, orderly and well-planned. These people plan well and don't over-commit. They do their taxes ahead of time, tidy up when their spirits need a lift and organize their closets routinely because that feels so good to them.

Others choose someone who also has ADD. The attraction is obvious; each understands the other. Two people who feel different and misunderstood in most of life's venues get a respite from that in their marriage.

Finally, some choose a drill sergeant for a spouse. This is someone who enforces order both in themselves and others. They impose organization on the ADD partner swiftly and effectively.

The strengths and challenges of each of these couplings are different. The ADD-executive secretary marriage can have a nice "balance" and work well in the world. One partner brings stability and order while the other keeps life lively and "spices" things up. One keeps things grounded and sane, the other pushes for change. One gives life roots, the other gives it wings. Complementarity is the strength of this type of marriage.

The challenge for these partners is to appreciate and value the vast differences in the other. Two people who are such polar opposites confront these differences at every little decision. One wants to balance the checkbook, one wants to empty it. One wants to tidy up, one wants to party down. Marriages between such different people can work if:

- both are deeply in love with and committed to each other,

- both are willing to negotiate the frequent differences,
- both value what the other brings to the relationship
- and both are willing to work to understand someone who is very different from themselves.

Marriage counseling is often helpful in developing the skills for this first relationship pattern.

Relationships in which both partners have ADD are quite different. Two people with ADD tend to have instant and intuitive connectedness. They think alike and often read each other's minds. They understand each other's struggles and feel empathy easily. It is easy to see how they fell in love; they are kindred spirits. They do spontaneous and interesting things, collect quirky things and can be quite entertaining. On the other hand, they tend to have messy lives, chaotic homes and struggles with productivity. Neither has the skill to balance the other's deficits, so they are dependent on others outside the marriage to bring order and stability. These marriages can work well if:

- both are deeply in love with and committed to one another,
- mutual weaknesses are openly identified and addressed
- and liberal use is made of coaching, expert advice and organizational help

The final type of relationship can be a difficult one. The operating premise of a drill sergeant is the incompetence of the recruit and the superiority of the officer. It happens that some people like to be in a relationship where they feel that they are helping to fix others. And some people with ADD feel that they must get help to be fixed. Giving up control to someone who is willing to assume it can appear to be a solution to the problem. The ADD partner can thrive initially with the outward structure that the other imposes.

The challenge of this type of relationship is that it may not accommodate the growth of the "subordinate" partner. Having other people direct and control us was appropriate in grade school; back then, it helped us get our stuff done. But adolescence began the delicate process of assuming responsibility for the organization and direction of our own lives.

Many of us struggled with that. The high dropout rate of people with ADD in college is due in part to "life failure", possibly even more than it is to actual academic failure. Many adolescents leave the structure of their childhood home and are distracted from their life purpose by drugs, travel, leisure, boredom, poor choices, etc. If a partner comes along who "rescues" us from this extended adolescence by imposing rules, order and organization, it can give the feeling of "arriving" as an adult and overcoming the adolescent crisis. But in truth, it doesn't. Letting others direct or even control us does not solve the adolescent crisis; it returns us to our childhood safety.

The challenge of this type of relationship is that the subordinated ADD partner will probably develop a desire for maturity at some point. It may happen after a few years of marriage or at a crisis time. If it hasn't happened already, it will almost certainly happen when ADD treatment begins. At this point, a marital crisis will likely begin. It can resolve well if both parties are willing to change roles significantly. The drill sergeant has to give up control and become a true partner. The subordinate partner needs to assume adult responsibilities and cut the dependencies. None of this is easy. Strongly consider professional help if you feel that you are in this type of relationship.

Now not every relationship fits these three patterns. Life is not remotely this simple. My hope is that this discussion has started you thinking about how your ADD affects your most intimate relationship[33]. Let's turn now to some more practical ideas for making ADD marriages work.

We need to work to decrease the impact of our ADD on those we love.

Whether or not we are now medicated, most of us would probably admit that our spouses are honest-to-goodness saints on earth. They have cleaned up behind us, educated us, put up with our foibles and forgiven us. In response, we have forgotten anniversaries, forgotten what they said, piled up bills, and neglected housework and parenting. Everyone's story is quite different, but the more we know about ADD, the more we realize that it has been a burden to the one we promised to love and honor. What follows is by no means an ADD marriage manual, but a few prompts that I hope will help improve our marriages.

- We are especially dependent on the forgiveness our spouses extend us. The proper response to forgiveness is gratitude. Find tangible ways to say thank you. Think about flowers, cards, love notes. Save chocolates for worse than average offenses. (We could cause a rash of obesity otherwise.)
- Putting up with us can be tiring. Figure out how to help. The most endearing thing we can do may be to empty the dishwasher or fold the laundry.
- Temper the spontaneity. Many practical spouses don't want to be surprised with a weekend getaway to New York City. We may be more than enough spontaneity for them already.
- Mundane things matter. The path to great marriages begins with picking up our underwear. Figure out how to do the things that matter to your spouse, even if it makes little sense to you.

---

[33] Please forgive the repetition, but Gina Pera has 384 pages of great marriage advice in *Is It You, Me, or Adult A.D.D.? Stopping the Roller Coaster When Someone You Love Has Attention Deficit Disorder.*

- Execution matters more than intention. Intending to do well is worth a little credit, but only for a short time. Marriages are long-term, trial–and-error relationships. If something that we intended well doesn't come across well to our spouse, it's not his/her fault. It's our job to do better next time based on what we learned this time. It is fine to say, "I'm sorry, I intended something different." It is not fine to expect our spouses to respond to our *intentions* time after time, if we are not progressing in our abilities to express the love our partners need. Let's say Sam who has ADD loves roses and his wife, Sarah, loves daisies even more. Sam will no doubt buy Sarah a couple dozen roses before he discovers that she has a real weak spot for daisies and actually only likes roses a little. Now maybe Sam dislikes daisies. Maybe he considers them cheap—not special enough to represent his great affection. Amazingly, many of the Sams among us will keep on buying roses, thinking that their Sarah is mistaken and will come around if we just buy her enough roses. Not amazingly, many of the Sarahs in our world will feel unappreciated when we buy them what we like, not what they like. Sam can buy himself roses, but for Sarah, he needs to buy the daisies.
- We are good at abandoning things that aren't working and starting something new. This is not a helpful skill in relationships. Marriages fail when the people in them fail to work through a hundred small impasses. They succeed when we approach every impasse, large or small, with the assumption that we *will* work through it, no matter what. If that sounds like a lot of trouble, let me suggest that it is a molehill compared to the mountain of effort it takes to untangle two lives, find another willing partner and start over again at the beginning.

Allow me to repeat the last point for anyone who hasn't yet succeeded in maintaining a long-term, healthy relationship. Those of us with a high need for novelty struggle when routine and repetition replace the thrill of new love. There are certain skills needed to maintain intimacy, and they are not necessarily innate. They *can* be learned, and counseling is often the best place to begin. Medication for ADD activates the brain regions responsible for self-control when there's no emotional intensity available upon which to draw.

And finally, a couple pieces of advice for non-ADD spouses:

- People with ADD can be clumsy at expressing love. You will probably have to teach us how to do the things that feel loving to you. Don't be shy to teach and be liberal with encouragement. But please be generous in understanding that most of us are filled with intentions to do good and hampered by poor ability to accomplish it. In several ways, that is better than a smaller love that is well executed.
- When your spouse is treated for ADD, you may not see the same progress he/she sees. You might see more, you might see less. Neither of you is wrong. Both you and your spouse will have valid insights into how the treated partner is progressing. Share them for what they are worth. Together, the different perspectives can give your physician and therapist a more robust idea of how things are progressing than either individual's opinions could.
- You no doubt understand the importance of order and organization. You may be tired of spontaneity. But change and excitement are as vital as oxygen to people with ADD. Injecting some of your own spontaneity into the relationship when you can is very affirming to an ADD partner.

Once again, this is not a "relationship manual". I hope it has spurred our thinking about how we interact with the people around us and how they respond to us. It has for me. I'm going to order some daisies right now.

Chapter 11 WORKING AT OUR BEST POTENTIAL

Job selection is an essential consideration for everyone with ADD.

If our original career selection was made witout consideration of our personal strengths *and* our ADD limitations, we might consider a new job. Notice I didn't say "...*would* consider a new job," but only *"might"*. Not all of us need new jobs. Some are already in great jobs. Some are in jobs that can be tweaked to suit us better. Others are in disastrously unsuitable jobs. What's the difference? What makes a great job? The answer to that question is complicated, since it is different for every person. For the sake of brevity, we're only going to talk about some general principles of job selection that relate to ADD.

Are you in the right kind of job already?

A few lucky people are still trying to figure out why this chapter is even here. You love your job, and you're great at it. In truth, there are ADD people who have done an excellent job of career selection. There are jobs that are good for people with ADD and jobs that aren't. If you've got one of the former, keep it if you can.

But if you are one of the lucky few who has a great job, don't rush off to the next chapter on family yet. It's important to understand what makes any job a good one (or should I say, an ADD-friendly one?) Otherwise, you might accept a promotion into a bad one. It happens more than you might think.

Some people with ADD achieve amazing economic success[34]:

- David Neeleman, the founder of JetBlue Airlines
- Paul Orfalea, founder of Kinko's
- Terry Bradshaw, former quarterback for the Steelers and now an entertaining commentator for Fox Sports
- Michael Phelps, Olympic gold-medal record swimmer
- Ty Pennington, star of Extreme Makeover: Home Edition

These represent a handful of people who leapfrog into prominent or leadership positions despite their ADD and are thereafter able to continue succeeding, in part because they have the means to hire others to complement and balance their weaknesses. The ability of a person with ADD to attain this degree of success generally depends on a combination of drive to succeed, high intelligence, novel insight, unique skills and good fortune. Unfortunately, stories of such occupational successes are often paraded in front of us to suggest that all of us with ADD could achieve success in the job market, if we would just try harder[35]. The real issue is whether ADD prevents many others from doing the same. And, of course, it does.

I love seeing human interest news stories about people who have triumphed over difficult circumstances to achieve success. Doctors told Wilma Rudolph's parents that she would never walk again when she was a child after an infection caused neurological damage, but through her strong will and dogged persistence, she eventually became one of the pre-eminent female track stars of the 20th century. However, that's a wonderful story precisely because it is so rare. It doesn't imply

[34] All have publicly acknowledged having ADD.

[35] More insidiously, these examples are sometimes used to suggest that ADD is not really a disorder or at least not a disabling one.

that the thousands who remain crippled from similar circumstances are lazy and need to try harder.

The half million-dollar loss in wages from ADD.

The majority of ADD adults have had a really rough time in the job market with numerous voluntary job changes and maybe a few involuntary ones, too. Research shows that we change jobs more frequently than those without ADD and get fired more often. Unemployment rates vary with the economy, of course, but at any given time, about half of those with ADD are unemployed. This is not just a little trend we're addressing, but a huge issue for most of us. The researchers at Harvard who study us found that we earn $10,000 per year less than folks without ADD. That's a one-half million dollar lifetime decrease in earning power[36].

The key to understanding why some with ADD succeed in the job market and so many others do not requires an understanding of a phenomenon called 'compensation.' (No, not 'compensation' as in how much you are paid.) The term refers to coping strategies—our ability to make up for a performance deficiency by devising an alternate means of accomplishing the same task.

There is a difference between compensations for ADD and treatments for ADD.

The distinction between treatments and compensations may be clearer if we discuss how it applies to other disabilities. Glasses are a treatment for near-sightedness. They return the vision to normal so that the near-sighted person has nearly the same

---

[36] The researchers concluded that the economic impact of just this aspect of adult ADD totaled 70 billion dollars annually in the US. There are other economic impacts of ADD including the cost to the government of unemployment benefits, mental health costs, costs of treating addiction, costs of damage repair in motor vehicle and industrial accidents and penal system costs for the increased rate of incarceration. No one has attempted to assess the full economic impact of adult ADD. For comparison, the American Heart Association estimates the annual economic impact of heart disease (coronary artery disease) at 50 billion dollars annually in the US.

visual abilities as someone with normal eyesight. Someone who was near-sighted before glasses were widely available might have become a watch-maker or a gunsmith where close work was all that was needed to do the job. You certainly wouldn't hire that person if you were interviewing lookouts for the *Titanic*, but you might find him a job arranging flowers on the first-class deck. You haven't improved his eye-sight one iota, but you have found a job for him where his impairment isn't a performance limitation.

The problem with compensations is that each one is only helpful in a few select circumstances. Our florist may be perfectly happy with his job and unconcerned about his visual limitations on most days. "What do I need glasses for? I'm getting along fine without them." But what if his dream is to be the ship's navigator, and he cannot see the instruments clearly? He will not qualify for that job. Or what if one night he needs to find his way to the life-boats and cannot. Then his visual impairment becomes tragic.

Glasses for the near-sighted are an example of a treatment, not a compensation. Someone with glasses sees nearly as well as those with normal vision and can choose between lookout, florist or navigator based on talent, education and desire. He could even aspire to become a ship's captain, a choice that is just not open for someone who cannot see distances well. Compensations are not so broadly effective. They help, but still leave a narrowed menu of options. Someone near-sighted without glasses has less choice. He can, at least, find one job, but he better like arranging flowers, because he's not going to be able to qualify for any of those other jobs in this example.

We would say that Marlin Shirley, the Olympic champion amputee sprinter, had compensated well for his disability if he had become a great arm wrestler or professional angler. But if he is *treated* with an artificial leg fitted properly, he can choose almost any sport he desires. He has far less restriction with a good prosthesis in place.

Compensation is a skill that accounts for the success that some have found in the job market even prior to treatment.

Some of us have found successful compensations for ADD in the work world, but others haven't. There are three basic types of compensation. First, one can find a niche in the world where the things we are worst at aren't necessary. Imagine a job without deadlines or a clock to punch. Second, one can find a way to make up for what we are not good at doing—a "work-around" or a "crutch", if you will. In the old days, this was called a secretary. Third, one can find a job that isn't impaired if you actually have ADD. Think Alpine ski racing, cliff diving or writing books about what it's like to have ADD.

What is better—compensations for ADD or treatments? At first, you might say that the answer is treatments. In our example of the near-sighted flower guy, it is clear that "compensation" limits him to the very few places that he can function, whereas "treatment" (i.e. glasses) opens up many new potential career paths for him. Treatment in this instance is clearly the superior approach. But, it's not that simple with ADD. Both are actually important to employ. Finding the right niche is how we work up to potential. (This will become clearer as you keep reading.) Treating ADD improves our ability to do the repetitive and mundane work inherent in every job. Compensations are the path to excellence, and treatment is the way to avoid failure. Both are essential for real success. We've been discussing treatment for most of the previous chapters, so let's look at these compensation methods more closely.

Compensation method #1—the protected niche. Find a job where fewer organization and attention skills are needed.

There are still a few positions available where precision, multi-tasking and organization aren't needed. Someone has to man those little shacks on the fishing docks where they sell bait, right? There are a few really laid-back jobs in the world that involve very little paperwork, where showing up is the hardest part of the day. But there aren't very many, and most pay well below minimum wage, because they tend to be in developing countries, not anywhere near your home.

In order to illustrate this concept, imagine a school of salmon migrating upstream to spawn. One fish is missing a fin so it doesn't swim as well as the others and is really struggling to continue upstream. It finds a little protected cove where there is very little current and hangs out there. Any person looking in on that salmon would say, "See, that is a perfectly healthy salmon swimming about in this river perfectly well. The missing fin doesn't disable it at all. How else could it have come this far if it was disabled?" The person is right in a small way. True, the salmon isn't drowning or decomposing, so he is "healthy" in a limited sense. But he isn't swimming upriver anymore which is what typical salmon do. He has reached the limit of his ability to compensate and so is going to drop out of the great salmon migration to the spawning beds.

Since most jobs require varying degrees of paperwork and organization, many with ADD find our niche in jobs that lie well below our potential abilities. We drop out of the "jobs race" and settle for a quiet little protected pool by the edge of the job market. It's astounding how many ADD people with extremely high IQs drop out of college and find work in factories, small specialty stores, manual craft shops and the like. This lets them make a living in lower-skilled jobs that involve less planning, organization, responsibility and paperwork than the management and executive level jobs one might expect them to attain. They become employees in businesses that we might expect them to own or to manage. They have settled in a quiet pool and quit swimming upstream[37].

[37] You know, there is nothing at all wrong with picking a low stress job. My concern is that people with ADD who have the brains to run the shop often settle for working in it, because their inattention makes it virtually impossible to do the work that they are otherwise capable of doing. If someone who is perfectly capable of excelling in either the worker's role or the owner/manager's role freely chooses the lower income of the worker's role for personal reasons (such as to gain the fixed,

Jeff was diagnosed with ADD in his 40's. His marriage had just ended and he was depressed. He was trained as a nurse, but had left the field and worked in a string of sales jobs, all with limited success, much frustration and little satisfaction. He dreamed of becoming a nurse practitioner, but didn't dare return to school, because he had done so poorly there in his 20's, struggling to maintain the required C average. Within a few months of beginning medication and counseling for his ADD, he returned to school in the evening while maintaining his daytime sales position. For the first time in his life, he earned straight A's and made the Dean's List every semester. His sales work was also doing so well that his income rose to a level that make him think twice about changing careers. He considered staying with his sales work, but only briefly, because the satisfaction of success in his nurse practitioner work really did fulfill a lifelong dream. In retrospect, Jeff understood that the years of sales work—for him—represented a compromise of his life goals that he had made due to the burden of ADD symptoms. It was a detour that ADD necessitated[38].

Curiously, some high-powered positions can be protected niches, too. A good secretary can make a scattered, unfocused professional seem much more organized and competent than he really is. Both my patients and I certainly owe a great deal to all the nurses and office workers over the years who have kept me organized and on-task.

---

limited hours), I am happy for him or her. I am just as concerned when a physical disability causes someone who loves manual work to accept a management role. The point in every case is to treat the disability—whenever possible—so that it does not restrict job choices.

[38] Stories such as this invite a frequent question: Why shouldn't everybody—ADD or not—take ADD medications so that everyone can get a better job, be happier and fulfill their potential? The answer is that most people are not disabled by distraction and lack of self-control as Jeff was. His medicine doesn't make him smarter, but it removes barriers that impair his learning process and his task execution. He is very smart, very principled and has tried many ways to improve himself over the years, even though these compensations accomplished very little. He continues these efforts now that he is on medicine, but enjoys more improvement for his efforts than he has experienced before. People without ADD have little or no functional improvement on the same medicines.

Compensation #2—workarounds. Find another way to get things done.

There is more than one way to skin a cat, and there is more than one way for us to do the things people with ADD don't do well. I know quite a few physicians with ADD. Medical diagnosis is something that ADD brains can often do quite well, but there are mountains of paperwork that doctors must do, too. Even with excellent secretaries and nurses helping, there are still big stacks of paper that a physician must review on a daily basis. The common workaround that I see physicians using is time. ADD physicians generally don't let down their standards of patient care, they just use more time to get the same amount of work done. They either see fewer patients than their partners or get home later (or both). Happily for their patients, I don't see these physicians compromising professional standards to save time. Sadly, it's the doctors' families that pay the price for their inefficiency. Excess time spent at work is a very common workaround for people in many fields, not just medicine. It is more often used by people in salaried positions than hourly workers. People with ADD who do hourly work sometimes take part-time work or job share with others to reduce the time pressure.

Another example of a workaround is outsourcing—getting others to work for us. It can be as simple as paying our bills automatically (assign the work to our computer) or hiring a secretary, personal assistant or clean-up person to do the part of our job that we struggle with the most. Hiring an ADD coach can be a way to 'outsource' the management of our professional goals. Even for people working in positions where hiring another is not an option, there are forms of outsourcing that can be arranged. One worker can barter with another within their job descriptions. For example, one lawn maintenance worker could volunteer to load tools onto the truck if his counterpart would only prepare the dreaded customer-billing paperwork. These are small examples of literally thousands of workarounds that people with ADD have employed to help get work done.

Compensation #3—high intensity jobs—find work that people with ADD do well.

This is a form of compensation that only works for a select few with ADD. The inattentive ADD symptoms don't lend themselves to any jobs that I know of, but hyperactivity and impulsivity don't necessarily cause problems in a few situations. I do a lot of traveling and lecturing in my work and will often point out that I have a dream job for someone with ADD—I am actually paid to talk a lot!

The risk-taking aspect of ADD is not present in all of us, but for some who possess it, there is a world of extreme endeavors that can be very satisfying. There are people who are *paid* to snowboard, cycle, parachute, climb mountains, shoot rapids and perform daredevil stunts. Can you believe how lucky these people are? Most of us only get to do these things on vacation.[39]

Another occupation that I would put in this class is overseas missionary work. It takes a special kind of person to leave one's homeland, learn a new language, adjust to a new culture and start a new organization in an underserved place. You'd almost hate to send someone overseas who doesn't have ADD if you think about it.

Compensation #4—finding our passions—the most effective compensation method of them all.

There is one thing that people with ADD do *without any impairment*. When we are involved with something we feel passion for, no better worker can be found. We can literally give 110% when we are emotionally invested in our work[40]. A

[39] You can tell whether you have the risk-taking gene or not by your reaction when you read that list of activities. If you sighed wistfully and privately envied the people who get to do those things all the time, you have the risk-taking gene. If your mind was drifting away with disinterest, you don't.

[40] Actually, everyone, ADD or not, can give 110% effort when emotionally engaged. It's important to recognize that we are *not* very impaired relative to others in the things we do fueled by passion, but we *are* impaired in the routine tasks. If tasks that are important on-the-job are not compelling to us, we tend to underperform.

few skilled managers know how to engage our passions in most any task, although they are unusual. But if we are involved in something we love to be doing, we don't have to wait for the right supervisor to come along. We supply our own motivation.

At the opposite end of the spectrum, it is especially sad to see people with ADD try to maintain a job for which they have no real heart or passion. Under these circumstances, we cannot perform consistently and have to live without the satisfaction that comes from work well done. When we perform badly we tend either to quit and move on or to be fired.

I can't tell you where your passion lies. Everybody's a bit different. Many of us have career paths that have centered on issues other than passion such as family expectations, subjects that interested us in school or the attraction of higher earnings in certain career paths. A few questions can help us begin to appreciate our deepest passions:

- What would you do with your life if money was no object and success was guaranteed?
- What have you enjoyed since you were little that you can't imagine living without?
- What were you doing at the happiest times you can recall?
- What do you find yourself doing when nothing else is pressing?

The answers to these questions don't tell you which job you should be doing, but certainly give you clues to where your heart lies.

And it's not as though we all have only one passion. I've had some wonderful hobbies and part-time interests over the years and have some others that I'll get to explore in coming years. Some smaller passions come and go. But learning and problem

---

Coworkers who don't have ADD can harness executive functions to do such routine tasks well in addition to the parts of the job about which they feel passionate.

solving have always been passions for me, so medical school was not an upstream effort; it felt to me like precisely what I was put here to do.

There are many types of jobs that can be ADD-friendly.

A job title all by itself does not guarantee that a job is right for someone with ADD. One forest fire smoke-jumper might have a job that is pure action and adventure, while another has to do mind numbing amounts of radio communication and logistics involving very little action. And no job is entirely ADD-friendly. We're going to try to select jobs that have large parts that we can do well and smaller parts that are troublesome. I love my lecture trips, but then there are expense reports to file if I want to actually get reimbursed. I do the whole trip out of enthusiasm and love of my work, but I get the expense reports done with a few workarounds and only when I'm on medication.

Everyone is going to have to do some slogging through the boring parts of the job. People without ADD have to do stuff they find boring, too. It might be easier for them, but they don't *like* to do paperwork either; they just don't hate it and fear it as much as we do.

Finally, the fact that a job is ADD-friendly doesn't mean it is a good fit for everyone with ADD. You might hate doing my lectures. Others might dislike sales. Fortunately, there is a wide variety of work available; let's look at it.

Examples of ADD-friendly jobs:

There are several general job categories that have a reasonable chance of being "ADD-friendly":

- Sales positions. People with ADD often do well in sales positions. These can work especially well if there is a "finisher" or "processor" that wraps up the paperwork. One example of this is real estate sales. The majority of time spent in home sales is prospecting and selling. Much of the paperwork is done by the banks, escrow agents and title companies.

- Entrepreneurs. Starting a company or starting a major initiative within a larger company can be just the novelty that it takes to engage our creative energies.
- Diagnostics. Jobs that involve figuring out problems are often well-suited to people with ADD. Intuitive thinking skills are the heart of these jobs. Such jobs range from physicians to auto mechanics to help-desk assistants to business consultants.
- Invention and product development. Working on the creative and cutting edge can fit well with ADD characteristics. Product design, testing and improvement are often about "invention" as well.
- Creative arts. Many with ADD have notable artistic and musical skills that beg to be used.
- Fast-paced work. Some thrive on the pace of a newsroom with its hard, frequent deadlines. Emergency work such as that done by paramedics, firefighters and ER nurses is a natural fit for some with ADD. I've diagnosed several people in this type of work who were recently promoted to desk jobs and suddenly figured out how much they missed the "action". They didn't know they had ADD when they were running like mad, but it sure became apparent when they needed to sit down and push paper. In retrospect, they had left jobs to which they were well-suited.
- Outdoor jobs. Park rangers, outdoor tour/trip guides, camp counselors, ski instructors, agricultural workers, groundskeepers and similar jobs allow active people to remain on-the-go throughout the work-day.

The point of this list is to show that many jobs have qualities that are "ADD-friendly". It is not by any means a complete list from which you must choose.

So, do you need a new job?

It comes down to how closely your current job matches your passions and abilities in life. If you are doing what you love doing, it's probably best to spend your efforts on developing compensations to help you improve your job skills. If your heart is not in your work, you *might* think about a change. But do it responsibly and carefully. We don't need to make impulsive decisions that are likely to go badly.

It took three years from the day I decided to devote my practice entirely to ADD until the day it actually happened. I had to endure some setbacks and delays along the way so that the timing could be right for my family and my former practice. It was definitely worth the effort and the wait—I love my current work. But job changes are risky and expensive affairs. They must only be attempted by professional drivers on a closed course. Wait!—wrong risky situation. They should only be attempted by professionally monitored individuals under a strict regimen of ADD treatment.

Now for some real expert advice on the topic of ADD and careers…

Wilma Felman is an expert on exactly this topic. Her book, *Choosing the Right Career,* is a much richer, more practical help to those who are not yet in a career that fits. If this chapter has helped you identify an issue you need to pursue, her book may be just the guidance you need.

Chapter 12 FAMILY MATTERS

ADD impacts families in profound ways.

Remember David Neeleman of JetBlue and his *60 Minutes* interview? The most telling comment from that interview may have been an off-hand remark he made. He said that he would not use medications because he feared losing his ability to think "out-of-the-box", but that his wife would probably be happier if he took them. I'll bet she would be happier. She might want to call my wife to get an idea of just how much happier. Then again, maybe she wouldn't want to hear it. It can be painful to imagine what you are missing, if there is little hope of change for the better.

Terry Bradshaw has faced these issues and relates with refreshing honesty what is really important in his life. Terry has ADD and speaks briefly about it in his laugh-out-loud autobiography, *It's Only A Game.* Throughout the book, he talks quite candidly about his three divorces, his failure to form lasting friendships with the teammates that he led to four Super Bowl titles as well as his strained relationship with his former coach, Chuck Knoll, whom he nonetheless admires and considers a great coach. There is obvious sadness in his words when he talks about his personal failures and most especially his two daughters with whom he has limited visitation since his divorce from their mother. In the preface to the book, he says:

> "To have had such success in my professional life and such failure in my personal life is difficult for me. I would gladly swap them even-up, gladly."

Sixty Minutes should have interviewed Terry Bradshaw to get ADD really straight. There are people who have professionally triumphed despite it. But career is not the most important thing in life. ADD generally impacts family life much more than careers.

Our family life can be one of the most important things that we will experience on this earth.

I believe it is wise to take a long view of our lives. Brian Tracy is a motivational speaker who urges people caught up in the moment to step back and ask themselves what an issue will look like ten years from now as they look back. Most of life's worries and fusses disappear when you do that. But I'm proposing an even longer perspective on this issue of family and ADD. What will you think at the end of your life as you look back at these years?

As a physician, I have been fortunate to be with quite a few people as they approached death. Watching these events has taught me some valuable lessons. Only what matters the very most in your life remains at the end. People with strong family ties are the happiest. People with wealth rarely mention it, but people who were successful often express pride in their accomplishments. People with strong faith have peace in those days. These lessons are an especially memorable benefit of my years as a family doctor.

Improving our family life should be one of the main goals of our ADD treatment efforts.

If long-term goals are what my life is *really* about, then clearing up a messy desk is not really as big a deal as cleaning up my messy relationships. I hope you feel similarly. Because if you do, we need to be quite specific about the treatment effects that we target. It is good to get more organized and less frazzled, procrastinate less, remember to follow through when we promise to do an errand and to control our impulsive spending.

But the *real* measure of successful treatment of our ADD is how we behave towards family and friends. Impulsive spending only reduces our net worth; impulsive thoughtless remarks leave wounds and scars. Misplacing our keys only makes us late; a pattern of reactive anger with our children can mar their self-confidence. When we remember that a friend could use cheering up but forget to follow through, both of us miss a chance to experience a small but wonderful thing.

When one of my patients returns after a medication trial and reports that he can focus better on his paperwork and complete it in less time, I am pleased. That is a good outcome. But when his wife says that he is getting home earlier from work and that their family life is improving, I'm happier yet. That is what I work for—to see relationships improve.

When we judge the effectiveness of our medicines, it is good to measure them in our workplace, but it is best to measure them in our homes. Our progress can sometimes be measured by improvements in punctuality, but more often it should be measured in terms of our relationships. How many "just thinking of you" emails did we actually send compared to how many we thought of sending? Are we getting better at listening to our spouse and empathizing when pain or disappointments occur? Can we rebound when relatives disappoint or annoy us? Are we consistently striving to give our children guidance and discipline? Are we better at working selflessly for our sons' and daughters' long-term welfare with every year that goes by? I hope so. These are the true measures of the treatments' effects.

Talking to family members about our diagnosis.

It's not always our first impulse to tell other people that we have ADD. Many people don't understand it or they still believe in some of the myths we have discussed. This can be especially sensitive in our extended families. There is often a fear that someone will ridicule us for admitting our weakness. People sometimes interpret sharing about ADD to be "making excuses" or "blaming a diagnosis" for our failures. More than one person has had a family member say something like, "So if *you* have ADD, then what does that say about your cousin and your uncle who are just like you but more so? Are you saying that they're mentally ill, too?"

If you know that there is hostility in your family toward the idea of ADD, you might want to wait to bring it up to relatives. You could wait until the treatment is helping you make visible progress in your life. Then you can at least point out concretely where the treatments are helping.

But I recommend that you keep moving toward a goal of discussing your ADD diagnosis with your relatives for a different reason. Your diagnosis is important information that they might need to act on themselves.

When doctors make a diagnosis of a condition that runs in families, they generally suggest that blood relatives be evaluated too. If someone in your family ever had colon cancer or even the benign colon polyps which can develop into colon cancer, they were instructed to share the fact of their diagnosis so you could be tested too. We see the same in many diseases such as diabetes, heart disease and breast cancer. If one of these is present in your family, your physician will be more aggressive about looking for it in you than in patients without that family history.

ADD runs in families with the same predictability that height does. If one of two parents is quite tall, there is an excellent chance that some or all of the children will be tall. If both are short, short children are likely. These results are not certain, but highly likely. The same high likelihood exists when ADD is present in a family. If one child has ADD, there is a 57% chance that one of the parents has it, too. If two children have it, that chance rises above 70%. If all five children in my family have it, and my mother doesn't, what is the chance that my dad has it? Let's just say that the odds are so high that he might as well skip testing, save the money and go straight to taking medication. Well, not quite, but you get the point.

What I hope you do with this information is straightforward. As soon as you feel comfortable telling close family members that you have ADD, you should do it. Urge them to learn more about it and consider testing if they have similar symptoms. Even if they don't have ADD, learning more about it might help them understand a few of your quirks. And you might gain a valuable ally who can help measure your progress and give valuable feedback. Maybe they already noticed that you no longer dominate the conversation at Thanksgiving dinner, but had no idea that it was due to your treatment. It might be just the first of many improvements that you make in your relationship with that relative over many years.

Family relationships are layered and complicated. This is not to suggest that all of the past history of factions and squabbles within a family will clear up just because we start treatment and become more thoughtful. Help from a family therapist is often the smartest next step we can take to help us restore and heal our families. The rifts that have opened over many years may require much work over many more years to repair.

But maybe, when that ultimate day of perspective arrives, there will be one more person at your final bedside, grateful for your life. I hope so. It's what I had in mind when I sat down to write this book.

Chapter 13 SPIRITUAL REDISCOVERY

Our oldest son, Ben, was eight when he learned the truth about Santa Claus. It was a pretty tough night for him. For almost two hours, my wife, Chris, and I consoled him as he cried and worked through all the instances of "deception" that had fooled him each year. ("You mean *Dad* was the one who ate the cookies I put out for Santa?" "So you wrapped the presents from Santa differently just so I wouldn't know it was *you* who wrapped them?" and so on.) Then, he started figuring out a couple other issues. "I'll bet you're the Easter Bunny, too, aren't you?" This went on for quite a while. Our hearts were breaking for him, and it crossed our minds that maybe the whole Santa thing isn't worth what it costs in the end.

At one point, I began to worry that he was going to lose all trust in anything that we had told him. So I asked him, "Does this make it hard for you to trust other things we've told you that you can't see, like God?" He looked at me with fire in his eyes. "No," he said with fierce conviction. "Jesus and God are the *only* things I believe in right now."

The implication was clear; he wasn't all that sure of how the world worked, why adults do what they do or how kids are supposed to survive childhood. I'm not sure he believed in *me* in that moment. But in this stormy sea, his anchor held.

It's time to talk about really central values in our lives.

What really makes our lives meaningful? What keeps us going when everything is falling apart? What centers us? What do we believe?

There is really no scientific data to help us in matters of faith and health. We are beyond good studies and diligent research. We're discussing Truth. As a physician, I have little right to go here. As a fellow traveler learning how to best live with ADD, I have no right to ignore it. If I had much sense, I'd leave sensitive and potentially divisive topics like this alone. Clearly, I'm not that sensible. I've got ADD…right? I stumble into all sorts of places where more sensible people know not to go. But, I've always felt that there should be more discussion of the intersection of faith and medicine, so I'm going to blunder in here as well.

I'm a Christian who has struggled through these issues. I've seen Jews, Muslims, Agnostics, Atheists and other Christians struggle through them in many different ways. But watching and doing are not the same, so I'll speak for myself in this chapter and hope that it prompts you to think for yourself. I'll be telling my own Protestant story. I hope that readers of different faiths will find helpful parallels despite the obvious differences.

I have noticed that many people with ADD are deeply spiritual, but not outwardly "religious".

Many of us with ADD don't appear very spiritual at first glance. Smoking, drinking, acquiring sexually transmitted diseases, divorcing spouses and losing jobs are not exactly what you do to impress your priest, but they are just some of the things that we with ADD do more than others. I'm not aware of any studies of church attendance in the ADD population, but I suspect it is lower than in the non-ADD population. We are over-represented in jails, AA meetings and unemployment lines, but probably not in the lines for communion.

But when I ask people with ADD about their faith, the responses are not what one might have suspected. Two themes have emerged. First, many people with ADD feel out-of-step with traditional faiths for reasons that have something to do with ADD. They were born into actively religious families but have left the practice of their faith. Feeling like a misfit is just as true for people with ADD on Sunday morning as it is every other day of the week.

Second, I am amazed at how many people with ADD have held tightly to their faith through all the failures and setbacks. Something strong and resilient inside of people with ADD keeps us bouncing back thousands of times from the same failures. How many times have we missed goals or fallen short of our own or others' expectations, yet we soldier on. The sad flip-side is that we have an impaired ability to learn from these mistakes and adjust our course to avoid making them again. But the miracle is that we try again. Like the salesman who doesn't mind getting turned down 50 times if that's what it takes to make one sale, people with ADD often have an amazing ability to put failure behind and try, try again. Some anchor is holding. Somewhere inside, even as our self-esteem erodes, something tells us that we have more potential and must keep going.

The sense of having a potential worth striving for should have naturally died early in people who have a long history of not living up to potential. But even when I meet 50, 60 or 70-year old individuals who are just being diagnosed with ADD (meaning that they have lived that many years without understanding their disability), there is always a remnant of self-respect and hope that lives inside. "Smarter" folk might have given up, but we press on. ADD people have been knocked down a bit more than others, but still have faith that we could be and always have been much more than what we seem to be.

My sense is that the "something inside" that anchors many of us is our faith in God. It's what got little Ben through that very hard discussion. It's what has gotten many of us through all of our days and especially the worst ones.

Did ADD treatment help improve my faith?

ADD can make it tough to be a Christian. Two major components of a typical worship service are the sermon and the prayers. Participation in either takes major concentration. Most of the Christians with ADD with whom I have talked feel guilty about how little they participate in traditional worship. Many have abandoned the regular practice of their faith, not because of disbelief, but because of a sense of being "out of place" in a worship service.

Before I was diagnosed with ADD, I thought that some patterns in my life represented sinfulness, and they caused a perpetual sense of shame in my life. Christians are called to be patient, and I am often impulsive. We are called to pray and read the Scriptures, and I almost always have trouble focusing on God. We are called to think of others and I am often self-centered. Christians are instructed to live in communities and love each other, but I am not very good at the friendship, intimacy and commitment that requires.

But something remarkable happened when I began medication for ADD. My patience improved, my prayer life got better and I could listen to sermons and remember them later in the week. I found myself more able to think about others and able to act in their interest with less regard for myself. It was hard to understand and characterize what was going on. Had a pill actually improved my morality, my spirituality?

C.S. Lewis, in *Mere Christianity,* wrote about the difference between mental health and morality. A person's morality has to do with the efforts made to do what is right. While mental health problems may hinder those efforts, God knows how much ability we possess to be good and expects us to exercise improvement beginning with what we are given. Improved mental health is simply more "raw material" available to do what is right. In other words, someone who is born with very little patience and who displays "all of it" is probably more morally advanced than someone who is born with much patience and exercises only some of it. She will probably appear less patient to those of us looking in on her life. But she is more obedient than the one who was given the gift of much patience and uses only a small part of it. Lewis goes on to say:

> "Humans judge one another by their external actions, God judges them by their moral choices…When a man who has been perverted from his youth and taught that cruelty is the

> right thing, does some tiny little kindness, or refrains from some cruelty he might have committed, and thereby, perhaps, risks being sneered at by his companions, he may in God's eyes, be doing more than you and I would do if we gave up life itself for a friend.[41]"

So there is no morality in a pill. Medications only give us a stronger foundation upon which to base our efforts to become better people—IF being better people is what we were seeking even before the pills came along. I suppose they will give you a stronger foundation upon which to become more cruel or dishonest, if that is what you hope for, too.

ADD treatment also helped me disentangle 'shame' and 'anxiety' from my experience of ADD.

Shame and anxiety are central themes in many recent discussions of the impact of ADD on the hearts and souls of people who have it. Both psychiatrists and theologians consider these to be important problems that invite intervention.

Anxiety can have a useful function in raising our adrenaline to deal with conditions that need extra focus and energy (e.g. wild animals nearby or an approaching deadline), but too much of it is considered a psychiatric disorder. Anxiety disorders are so named, in fact, because the excess anxiety interferes with and *decreases* our ability to respond to stressful situations. Theologically, anxiety is often seen to indicate a lack of trust in God. Most people recognize anxiety as unpleasant and will work and plan to avoid it.

But many with ADD seem attracted to anxiety. We procrastinate so reliably that we *must* not have the typical aversion to it. In my opinion, many people with ADD use anxiety as a productivity tool. We can *manufacture and use anxiety to be productive.* Procrastination is "scheduled anxiety". It helps us turn important tasks that have little emotional content ("Gee, these tax forms are boring.") into the emotionally compelling

---

[41] Lewis, CS, Mere Christianity, p. 71, 1943, Scribner

events to which we can respond. ("Gee, I'll bet I can still finish these tax forms before the post office closes.") My early study habits were designed around this dynamic before I knew I had ADD; I procrastinated to build up emotional energy to do the mentally challenging work of medical school. I never found another way to do that work.

Psychiatric shame is a sense of personal failure that injures self-esteem. Theological shame is a sense of failure that sinners feel when comparing our efforts to God's requirements. Pastors and psychiatrists both battle feelings of shame in their clients, but to us with ADD, shame can have a perversely useful function.

If we can't interest ourselves in a routine task, and if it doesn't have a deadline, we can still "shame ourselves" into starting the task by actively feeling the shame of failure, then working against it. Lacking the ability to do mundane things for importance sake, we shame ourselves into doing them. When we need to get our homes or garages organized, it's much easier to do if we can turn it into an emotionally compelling burden. We imagine that someone—our spouse, a parent, a visitor that might come—is disgusted with the chaos. We tear through the mess, furious at ourselves for waiting so long, chiding ourselves for the times we noticed the mess without starting to clean it, promising ourselves and God not to get in this mess again. Our counselors and pastors are working to free us from shame, but I wonder how many of them understand how useful and productive it can become for us.

The alternative for accomplishing important tasks if you don't have ADD is to simply do them. Executive function in the brain refers to the human brain's ability to plan, organize and carry out complex, goal-directed behavior when the reward is not immediately available. Those of us with ADD were born with diminished function in the areas of the brain that "simply do" mundane tasks in response to their importance. We have no diminished function in the areas of the brain that do what is interesting or compelling to us. Medications improve executive function so that we can execute tasks without invoking emotional workarounds to start and finish them. "Just do it"[42] becomes possible.

With treatment, we don't need anxiety and shame at every turn. We can respond to them and use their intrinsic energy when they are naturally present, but don't have to invoke them to be able to clean up a messy room or do our taxes on time. Those are "just do it" tasks. Prior to beginning treatment for ADD, I often imagined God as rigorous, demanding and disappointed in me. That wasn't my only experience of God (as you will see in the next section) but I turned many mundane tasks into religious obligations just so I could fear the failure to complete them.

When people say that they have decided not to seek treatment for ADD "because I've learned how to deal with it," they often mean that have harnessed the productive power of shame and anxiety to do what needs to be done. What a high price to pay, when there are therapies available that allow us to "just do it".

Conversely, my faith has helped heal my ADD.

There are several themes in my Christian faith that have been especially helpful to me in dealing with ADD (even during the 40 years before I knew I had it).

Christianity begins with the notion that God has high standards for how we should live, and that we have all failed his expectations. It's called "sin" in theological circles. Some of us have pretty high defenses and hate to admit we're ever wrong, but in honest moments, most people with ADD feel right at home with the notion that we are error-prone and fall short of what we should be.

The real magic of the Christian faith for me begins with the notion that God still loves me, despite my failures. This is where the healing effect of my faith on my ADD really begins.

---

[42] Nike's ad campaign using that slogan appealed directly to the same point. Exercise is not emotionally compelling most days, so we shouldn't wait until we *want* to do it. That day will never come. Do it now, because it's important. Brilliant!

Self-esteem is a tough issue for most everyone with ADD. Many of us are social misfits. Even those of us that are the "life of the party" find that there is a limit to how much of our levity other people want in their lives. Many are divorced—literally rejected by people who once promised they would love us until we died. The sense of loneliness that is so common for people with ADD arises from the failure to maintain close relationships over the years.

There is little I find more affirming than to be desired. It is an antidote to the expectation of rejection experienced by many of us with ADD. I am fortunate to have had parents who did a great job at instilling a healthy sense of self-worth in me from an early age, so I've suffered less than many others. The basis of their attitude was their firm belief that God made me uniquely for His own purpose. Despite my flaws, they always helped me feel I was someone special—not just special to them, but special to God as well.

Now, think what it would be like if your favorite current or recent president knew you by name. Imagine you're at a political gathering, and he spots you in the crowd of a thousand people and waves you to come over. He smiles broadly and introduces you to the other dignitaries. "Hey, everybody, I want you to meet [*insert your name*] from [*insert your hometown*], one of my favorite friends. I'm so glad you're here. This really makes my day." This would be a great story to tell back home. It's hard to imagine just how good that would feel.

But this isn't merely the president we're talking about; this is the God of the universe, the One who is so powerful that He knows seven billion people by name and cares immensely for each one. I grew up hearing that He knew my name and smiled when He thought of me. (Every Sunday of my childhood our worship service ended with the benediction: "The Lord bless and keep you. The Lord make His face to shine upon you and be gracious to you. The Lord turn His face toward you and give you peace." The language is old, but it's about God being happy to see us.)

Imagine what it does for me to think that God is happy to see me. He could be done with me. My behavior is not up to His standards. I'm not His type. However, for reasons that must have more to do with love than fairness, He wants me to come home in the end and live with Him forever. In my imagination, it goes something like this:

I arrive in heaven and go in to meet the *real* Supreme Court judge. It's awe inspiring and frightening—as if I was a disheveled backpacker stumbling into a palace and realizing how dirty and rumpled I am compared to the beauty and formality of the place. God is busy in some very serious task with some important archangels bustling about at his side. The gravity of this is overwhelming; my eternal destiny is in His hands. I'm not well dressed, and clearly feel that I don't belong. God looks up from His work and sees me standing there. A broad smile lights His face, and His whole demeanor changes. "Hey everybody, look who just arrived. This is Oren from Grand Rapids. You've got to come and meet him. I've been looking forward to having him here for quite a while now." And then a big party begins.

This image is a remarkably healing antidote whenever I struggle with self-esteem. In fact, growing up with this sense of how much God loves me is the reason I haven't had to struggle more with self-esteem.

Some other random thoughts about ADD and faith:

- The apostle Peter may have had ADD. He was often the first to blurt out an answer when Jesus asked questions. He hopped right out of the boat when he saw Jesus walking on water. Sounds pretty impulsive to me. It also seems impulsive that he attacked one of the Roman soldiers arresting Jesus in the Garden of Gethsemane and cut off his ear. A couple hours later, he lied three times when strangers asked him if knew Jesus. It is encouraging to me that one of the leaders of the early church was chosen despite his impetuousness.

- We've already noted that many missionaries have ADD. So do many pastors. Maybe so did some of the crazy outcasts like St. Francis, the Catholic saint. I am pretty sure that God doesn't need neurologically perfect people to get His work done.

- When I was 9 or 10, my Dad let me help in the church nursery instead of sitting through what I then thought was a stultifying service. That was a pretty good accommodation for me. Ever wonder if a certain pastor or choir director was drawn to the job partly because it doesn't require sitting still very often?

- Before I began ADD medication, I found it difficult not to perseverate on other people's failings and shortcomings. On medication, it is much easier for me to see others' points of view and to pay attention to my own business. This is probably what Jesus meant when he said to take the log out of our own eye before we try to remove the splinter from someone else's. It is most certainly a happier way to live.

- My friend, Gib, urged me to include a mention of meditation in this chapter. When I've tried it, the results are laughable; I've never emptied my mind for even a full second, or noticed any effect that could be called enlightenment. Gib has had more success than I, so I will try to remember to include his cell phone number in the next edition of this book in case you want to learn more.

All the great faiths may be therapeutic for us with ADD.

I know a fair bit about the Christian faith but too little about most others. As far as I can tell, many other faiths have much to offer us, too. Catholicism, Judaism, Orthodoxy, and Islam all have rich and ancient traditions. For some, the traditions provide a valuable anchor in a life otherwise ruled by a search for novelty. Conservative forms of all these faiths offer a structure of rituals that can work well to contain distractions for some. Others enjoy the novelty and innovation seen in more liberal settings. The Eastern religions such as Buddhism and Hinduism teach a mind-set of focused relaxation. Anxiety is not a problem to be solved, but an indication that we are too self-centered. Yoga and martial arts are generally divorced from their religious origins here in the West, but many with ADD have found them therapeutic for their ADD as well as helpful for balance, strength and flexibility. All of these have the capacity to help our ADD and paradoxically be even more fulfilling when our ADD is treated.

Some practical thoughts might apply

I hope these two points are clear. First, treating ADD may help us achieve spiritual goals. Second, our spiritual life may play a vital role in treating our ADD. Alright, so what's the "take-home" advice?

- If you are disillusioned regarding spiritual matters, ADD treatment might give you a new opportunity to re-experience your faith. Disillusionment may represent frustration over how ADD degraded the practice of your faith. Maybe you are a discouraged believer, not an unbeliever. If you have left a church or the practice of a faith because you did not "fit in", that might not mean you have lost your faith. Maybe it means that you feel left out or disconnected.

- Consider that some changes may help you find a more "ADD-friendly" worship experience. Several years ago, my family began attending a non-traditional church. It is multi-racial, located in a struggling inner-city neighborhood. There are many mixed-race couples in the church along with residents from a drug-rehab house, college students, immigrant families, suburban families and everybody else in between. Nobody is "normal", so anybody and everybody fits in. The music is lively and varied, the sermons are brief and thought-provoking, and the worship sequence changes every week, so that it is not predictable. This novelty factor is tremendously helpful for me, and, I suspect, would be for most people with ADD.

- Forgiveness is a central theme in most religions, and I believe it should be a central theme in our healing as well. We blame ourselves constantly; life is better when we learn to forgive ourselves. We blame others quickly; we need to learn to forgive them more easily. The practice of communion has been a wonderful part of my faith. The message of the service is: "God forgives you, so follow his example by forgiving yourself and others." It's been revolutionary for me.

- I'm at a loss to offer much to agnostics and atheists at this point, although I suspect some may have closed this book several pages ago. To those who haven't, kudos for your open-mindedness.

A parting note.

To everyone who does believe in God, even if it only seems a tiny and inconsequential part of you, I encourage you to return to your roots and re-examine your own spirituality. Life is hard, even after ADD is well-treated. By the time he was nine, my son, Ben, could already tell you how harsh this world can be, and how much we need an anchor.

Anchors are pretty small things compared to the boats they secure. The question is not how big our anchors are. The question is rather how solid is the rock to which they are affixed.

> *"May the Lord make His face to shine upon you and be gracious to you. The Lord turn His face toward you and give you peace. Amen"*

Chapter 14 PERSPECTIVES ON HEALING ADD

"A man down on earth needs our help." Joseph, the archangel was introducing the story of George Bailey, to Clarence Odbody, the new angel who was still trying to earn his wings.

"Is he sick?" asked Clarence.

"No, worse," replied Joseph. "He's discouraged."

--from "It's a Wonderful Life."

Healing ADD is hard for the impatient. It takes time.

The advice in this book is far from simple. We don't casually begin treatment and then assume life's highest functional level a few weeks later. It takes years to figure out and develop the new abilities that ADD treatments give us. And it will take the rest of our lives to grow along the new trajectory that ADD treatments make possible.

I believe that the first step that most of us should take after diagnosis is to try the medications available and select the best regimen we can find. This will take several months. The next steps involve lifestyle changes, education and consultation to start our lives moving in better, more satisfying ways. This may take a year or more. The long-term strategies involve partnerships, attitudes and beliefs to help us reach the potential we've always sensed we might. Setting these strategies in place may take a number of years

I hope that you are not overwhelmed by the length of the "to-do" list that this book seems to suggest. It takes time to implement all these recommendations, and it will take many more years to see all the good that comes from those efforts.

But there is one good thing that doesn't take so long—in fact, with some luck, it has already started. It is the development of the hope that we have found a way to reach our potential.

The first milestone in the healing of ADD is the appearance of hope.

Healing ADD begins when hope starts to replace discouragement. ADD is simply a brain dysfunction to a scientist, but its result is often a lifelong sense of failed potential for those who live with it. Frustration and discouragement are two of the most prominent themes I hear from my patients before we begin treatment. The main tasks of my first visit with a new patient involve diagnosis and treatment, but the real purpose is to kindle hope.

So my final wish for you is that you will find a renewed sense of hope, that you will see the potential for your life and feel hope instead of discouragement. I hope you are anticipating the fuller, more satisfying life that treatment can help you achieve. I hope you are remembering all your earliest dreams about what you would do with your life and who you would someday become. *Your* hope is the reason I sat down to write this book. Far more important though, I suspect it's the reason you finished it.

ACKNOWLEDGEMENTS

I've read the acknowledgement pages in many books over the years and wondered why so many people deserved thanks and wondered what they contributed. The author just typed up a manuscript and sent it to a publisher, right?

Wrong. You re-arrange your life to write a book, and—if you are a father and a husband—you only do that with your family's help and forgiveness. My own family has been incredible. My wife, Chris, is strong and beautiful and capable. She assumed my responsibilities so that I could have the time to do this. Her encouragement was vital; it's not just that she believes in me, but she has believed in this book and wanted to see it through because she believed it could help people. Her heart is large, and for some inexplicable reason—even though she is one of the least-ADD people I know—her life is full of people like me. She is a healing presence in so many ADD lives.

We still have two boys at home, and they have been terrific. Ben asked me to dedicate this book to him before the first chapter was written. "You wouldn't know very much about this if it wasn't for me, right?" But then everybody will know that you have ADD. "They already do, Dad." Paul told me I could use his name if it would help other people who needed it. He may be the most generous person I've ever met.

Several friends deserve special mention. Doug Bouman believed in this book more than I did some days. What a gift of encouragement he has! Alison Hodgson, Pete and Gayle Knibbe, Rick Osborn (who helped me sound like a much better writer than I really am), Barb Mann, Marianne Looney (who actually *went* to finishing school), Chris White, Connie Smith and Julie Thompson all helped with the difficult work of polishing the manuscript. Andrea Larson, one of my best friends ever since residency fixed more grammatical and stylistic mistakes than an editor should have to face. Dr. David Baron is a professional colleague who has been a friend and mentor since the day we met. His content edits were essential, but his encouragements have meant far more. Bill Oechsler of XLER Consulting tried to keep me "on message". Once I figure out what that means, I'll tell you whether he succeeded or not.

As I wrote this manuscript, I imagined many different future readers working through this book. You were all very helpful to me, and I hope we will have the chance to meet some day.

I could go on—my extended family and numerous colleagues come to mind. But you must know by now how very much the patients whose names I can't reveal have inspired me. Their stories of rediscovered potential are behind every page.

Thank you.

DISCLOSURES

Every "expert" has biases that affect his or her advice. We can't eliminate all bias and influence, but we can at least disclose them, so you can evaluate them for yourself. I have worked as a paid speaker and consultant for Eli Lilly and Shire Pharmaceuticals. I've also been paid to speak for or consult with physicians and professional medical organizations, school systems, teacher associations, independent research organizations, educational institutions and organizations, diagnostic clinics and patient advocacy groups. I don't hold any investments directly in any drug manufacturer or healthcare company. There, you have it…

AN INVITATION

It would be great to hear your feedback. What did the book say that was helpful, what needs work? There will be future revisions to keep the information in this book up-to-date, and I will seriously consider every thoughtful idea. Send emails to feedback@omason.com.

CPSIA information can be obtained at www.ICGtesting.com
Printed in the USA
269350BV00001B/1/P